Series: "Modern Mathematics for Engineers"

Tamara G. Stryzhak

Grassmann Algebra
and
Determinant theory

Grassmann Algebra

$$e_m \cdot e_k = -e_k \cdot e_m \rightarrow e_m^2 = 0$$

Tamara G. Stryzhak

GRASSMANN ALGEBRA
AND DETERMINANT THEORY

ibidem-Verlag

Stuttgart

Bibliografische Information der Deutschen Nationalbibliothek
Die Deutsche Nationalbibliothek verzeichnet diese Publikation in der
Deutschen Nationalbibliografie; detaillierte bibliografische Daten sind im
Internet über http://dnb.d-nb.de abrufbar.

Bibliographic information published by the Deutsche Nationalbibliothek
Die Deutsche Nationalbibliothek lists this publication in the Deutsche Nationalbibliografie;
detailed bibliographic data are available in the Internet at http://dnb.d-nb.de.

∞

Gedruckt auf alterungsbeständigem, säurefreien Papier
Printed on acid-free paper

ISBN-10: 3-8382-0089-6
ISBN-13: 978-3-8382-0089-7

© *ibidem*-Verlag
Stuttgart 2010

Printed in Germany

УДК 517.442.(075.8)
ББК В 161я7
А 62

This brochure presents the Determinant theory with the use of hypercomplex Grassmann figures. It enables to simplify the proof of many determinant features. It also states several ways of solving linear algebraic systems, most of which are connected to the Determinant theory.

The teaching experience in technical universities has demonstrated that the suggested way of presenting the Determinant theory is understood more easily.

The title of the lectures is designed to attract the readers' attention to a new and convenient form of presenting determinant theory for technical institutions.

All your opinions, remarks as well as advice and recommendations will be taken into account to improve the next editions.

Phone: +38044-406-83-48
Fax: +38044-406-82-20
E-mail: stri@aer.ntu-kpi.kiev.ua
Web-site: www.iaeste.org.ua

Dedicated
to the 200th
anniversary
of Hermann
Grassmann's birth
(1809 – 1877)

Hermann Grassmann (1809-1877)

Series: "Modern Mathematics for Engineers"

Lectures for the trainee-students of IAESTE

Educational publication

Grassmann Algebra
and
Determinant theory

This book contains the Determinant theory which uses hyper-complex Grassmann numbers to simplify the proof of a large variety of determinant properties. The book also contains some methods for solving linear algebraic systems. Teaching this material has demonstrated that the suggested method of presenting determinant theory is comprehended more easily by students and gives a more profound knowledge of the subject. The title of the lectures is designed to attract the readers' attention to a new and convenient form of presenting determinant theory for technical institutions.

Lectures
Professor NTUU "KPI" Tamara Stryzhak

Translator
Faculty member NTUU "KPI" Nataliya Sarycheva

English language editors

IAESTE trainee students:

Silbermayr Lena (Technical University of Vienna, Austria)
Paterson Colin (University of Strathclyde, Great Britain)
Rudolph Benedikt (Technical University of Munich, Germany)

"Modern Mathematics for Engineers"

Chief Editor
Natalia Konovalova

Translator
Nataliya Sarycheva

The Author
Tamara Stryzhak

IAESTE trainee students
Editors

Benedikt Rudolph, Technical University of Munich, Mathematics
Lena Silbermayr, Technical University of Vienna, Mathematics
Colin Paterson, University of Strathclude, Mathematics

Photographer
Alex Babynin

Contents

§1. Grassmann Numbers

In the 19th century attempts were made to generalise the idea of complex numbers. These efforts led to the creation of new types of algebra. In his work "The Study of Extended Values" the German mathematician Grassmann developed the foundations of geometric representation.

Grassmann started with a basis of a vector space $e_1, e_2, \dots e_n$, which he called first order values. Linear combinations of these units with real coefficients are called linear forms

$$X = x_1 e_1 + \dots + x_n e_n,$$
$$Y = y_1 e_1 + \dots + y_n e_n,$$
$$Z = z_1 e_1 + \dots + z_n e_n,$$

where $x_i, y_i, z_i \in \Re, i = 1, \dots, n$.

We let capital letters denote these linear forms, while lower case letters denote coefficients. Supposing $\lambda, \mu \in \Re$, then

$$\lambda X = (\lambda a_1) e_1 + \dots + (\lambda a_n) e_n.$$

$$(1.1)$$

The sum of the linear forms X and Y is the linear form

$$X + Y = (x_1 + y_1)e_1 + \ldots + (x_n + y_n)e_n .$$

$$(1.2)$$

Therefore the linear forms form a vector space and the following axioms are satisfied:

1. $(X + Y) + Z = X + (Y + Z)$ (Associative Addition);
2. $X + Y = Y + X$ (Commutative Addition);
3. There is such a zero linear form 0 when $0X = 0$;
4. For every X there exists a $-X$ such that $X + (-X) = 0$;
5. $(\lambda + \mu)X = \lambda X + \mu X$ (Distributive Scalar Multiplication I);
6. $\lambda(X + Y) = \lambda X + \lambda Y$ (Distributive Scalar Multiplication II);
7. $(\lambda\mu)X = \lambda(\mu X)$ (Associative Scalar Multiplication);
8. $1X = X$ (Identity).

Grassmann adds to this vector space structure a law of multiplication between linear forms that satisfies the following axioms.

1. $e_i \cdot e_j = -e_j \cdot e_i$

(1.3)

(Skew Symmetric Multiplication);

2. $\left(e_i \cdot e_j\right) \cdot e_k = e_i \cdot \left(e_j \cdot e_k\right)$ (Associative Multiplication);

3. $e_i \cdot \left(e_j + e_k\right) = e_i \cdot e_j + e_i \cdot e_k$ (Distributive Multiplication).

The result of multiplying two first order elements is a second order element, with basis e_{sk} where k and s take values from 1 to n. It is left to the reader to check that the span of this second order basis with real coefficients satisfies the above conditions and therefore also forms a vector space.

From (1.3) we see that the product of the equal elements is zero, that is

$$e_i \cdot e_i = 0, \qquad i = 1,\dots,n.$$

(1.4)

This can be proven from (1.3). Consider when $j = i$;

$$e_i \cdot e_j + e_j \cdot e_i = 0 \Rightarrow 2(e_i \cdot e_i) = 0 \Rightarrow e_i \cdot e_i = 0.$$

There exist C_n^2 different second order basis elements:

$$e_i \cdot e_j = e_{ij}.$$

We can form a third order basis by multiplication as follows:

$$e_i \cdot e_j \cdot e_k = e_{ijk}.$$

It follows from the condition (1.4) that an element e_{ijk} is equal to zero if the value of two of the indices i, j, k coincide. If the indices are rearranged, the value of the element e_{ijk} will change by a factor of -1. Thus by rearranging we find the following equality

$$e_{ijk} = -e_{ikj} = e_{jki} = -e_{jik} = e_{kij} = -e_{kji}.$$

It is left to the reader to check that linear combinations of third order basis elements with real coefficients also form a vector space. We may define the units of the 4th and the 5th orders and so on inductively. A trivial counting exercise reveals that there are C_n^r elements in an rth order basis. Each element of the rth order basis is the result of multiplication of r first order basis elements

$$e_1 \cdot ... \cdot e_r = e_{1...r}$$

and there do not exist any elements of higher order than n. Linear forms are also called Grassmann numbers. In general the multiplication of Grassmann numbers is associative and distributive, these properties are inherited from the basis. In particular, the following equalities hold

$$e_i \cdot \left(\lambda e_j + \mu e_k\right) = \lambda\left(e_i \cdot e_j\right) + \mu\left(e_i \cdot e_k\right),$$

$$\left(\lambda e_j + \mu e_k\right) \cdot e_i = \lambda\left(e_j \cdot e_i\right) + \mu\left(e_k \cdot e_i\right) = -e_i \cdot \left(\lambda e_j + \mu e_k\right),$$

$$\left(\lambda e_j\right) \cdot \left(\mu e_k\right) = e_j \cdot \left(\lambda \mu e_k\right) = \left(\lambda \mu e_j\right) \cdot e_k = \left(\lambda \mu\right) e_j \cdot e_k,$$

$$e_i \cdot \left(e_j \cdot e_k\right) = \left(e_i \cdot e_j\right) \cdot e_k = e_i \cdot e_j \cdot e_k = e_{ijk}.$$

Consider the general linear forms

$$X_i = a_{i1}e_1 + \ldots + a_{in}e_n, \quad i = 1, \ldots, n.$$

They have the properties (1.1) – (1.3) and so the linear forms X_i inherit the same properties as the first order basis.

Example 1: Find the product of the two linear forms given below.

$$X_1 = 2e_1 - e_2 \qquad X_2 = e_2 + 3e_3$$

$$X_1 X_2 = 2e_1 e_2 + 6e_1 e_3 - 3e_2 e_3$$

$$X_2 X_1 = 2e_2 e_1 + 6e_3 e_1 - 3e_3 e_2 = -X_1 X_2$$

Note 1. Using modern terminology we say that Grassmann calculus is an associative algebra over the field of real numbers, which is spanned by the basis $e_1,\ e_2, ..., \ e_n$ and defined by the relation $e_i e_j = -e_j e_i$.

§2. Definitions

Consider a square n×n array of the real num-

bers. We call this a square matrix and define a general matrix A as

$$
A = \begin{pmatrix}
a_{11} & a_{12} & \cdots & a_{1n} \\
a_{21} & a_{22} & \cdots & a_{2n} \\
\cdots & \cdots & \cdots & \cdots \\
a_{n1} & a_{n2} & \cdots & a_{nn}
\end{pmatrix} = \left(a_{ij} \right).
$$

(2.1)

The numbers a_{ij} are called matrix elements. The horizontal lines are called rows and the vertical lines are called columns. The number of lines (columns) of general matrix A is called the order of the matrix. Note that in the general matrix given by (2.1) the first index i of the element a_{ij} denotes the row number and the second index j denotes the column number. The line of elements a_{ii} with equal index values is called the diagonal. The other line of elements a_{ij} : $i+j=n+1$ is called the secondary diagonal.

Example 2. Consider the 2×2 square matrix

$$A = \begin{pmatrix} 2 & -3 \\ 4 & -5 \end{pmatrix},$$

where $a_{11} = 2, a_{12} = -3, a_{21} = 4, a_{22} = -5$.

We introduce linear forms associated to the rows of the matrix A

$$X_i = \sum_{j=1}^{n} a_{ij} e_j = a_{i1} e_1 + \ldots + a_{in} e_n.$$

$$(2.2)$$

Consider the product of linear forms $X_1 X_2 \ldots X_n$. This multiplication involves n^n terms. Some terms equal zero since they contain multiple factors of Grassmann units. Some components will be non-zero due to the fact that they consist of the product of different units $e_1, e_2, \ldots e_n$. Rearranging the adjacent multipliers and considering the property (1.3) we can make the product of different Grassmann units take the form $\pm e_1 \ldots e_n$. Finally after reduction of all the terms in this manner we arrive at the equality

$$X_1 X_2 \ldots X_n = D e_1 \ldots e_n.$$

$$(2.3)$$

The number D is called the determinant of matrix A and is defined as

$$D = \det A = \begin{vmatrix} a_{11} & a_{12} & \cdots & a_{1n} \\ a_{21} & a_{22} & \cdots & a_{2n} \\ \cdots & \cdots & \cdots & \cdots \\ a_{n1} & a_{n2} & \cdots & a_{nn} \end{vmatrix} = \left| a_{is} \right|_1^n.$$

The order of this determinant is n. Elements $a_{i1}, a_{i2}, \ldots, a_{in}$ create the nth row of the determinant and the elements $a_{1s}, a_{2s}, \ldots, a_{ns}$ constitute the main diagonal of order s. We shall consider the multiplication of the two linear forms:

$$X_1 = a_{11}e_1 + a_{12}e_2, \qquad X_2 = a_{21}e_1 + a_{22}e_2.$$

For the products $X_1 X_2$, using properties (1.3) and (1.4), we have

$$X_1 X_2 = \left(a_{11}e_1 + a_{12}e_2 \right)\left(a_{21}e_1 + a_{22}e_2 \right)$$
$$= a_{11}a_{21}e_1 e_1 + a_{11}a_{22}e_1 e_2 + a_{12}a_{21}e_2 e_1 + a_{12}a_{22}e_2 e_2$$
$$= \left(a_{11}a_{22} - a_{12}a_{21} \right)e_1 e_2.$$

Finally we find the expression of the determinant in terms of its elements

$$\det \begin{pmatrix} a_{11} & a_{12} \\ a_{21} & a_{22} \end{pmatrix} = \begin{vmatrix} a_{11} & a_{12} \\ a_{21} & a_{22} \end{vmatrix} = a_{11}a_{22} - a_{12}a_{21}.$$

We can express the determinant of a third order matrix in terms of its elements. To do this we consider the product of the three linear forms:

$$X_1 = a_{11}e_1 + a_{12}e_2 + a_{13}e_3,$$

$$X_2 = a_{21}e_1 + a_{22}e_2 + a_{23}e_3,$$

$$X_3 = a_{31}e_1 + a_{32}e_2 + a_{33}e_3.$$

While calculating the product $X_1X_2X_3$ the majority of the terms are equal to zero since they contain a product of equal units. So we shall only write down those terms which contain the product of the different units e_1, e_2, e_3. Thus we have

$$X_1X_2X_3 = a_{11}a_{22}a_{33}e_1e_2e_3 + a_{11}a_{23}a_{32}e_1e_3e_2$$

$$+a_{13}a_{21}a_{33}e_2e_1e_3 + a_{12}a_{23}a_{31}e_2e_3e_1$$

$$+a_{13}a_{21}a_{32}e_3e_1e_2 + a_{13}a_{22}a_{31}e_3e_2e_1.$$

Using the following equalities

$$e_1e_2e_3 = -e_1e_3e_2 = e_2e_3e_1 = -e_2e_1e_3 = e_3e_1e_2 = -e_3e_2e_1,$$

we find the determinant of a third order matrix in terms of its elements:

$$(2.6) \quad \begin{vmatrix} a_{11} & a_{12} & a_{13} \\ a_{21} & a_{22} & a_{23} \\ a_{31} & a_{32} & a_{33} \end{vmatrix} = a_{11}a_{22}a_{33} + a_{12}a_{23}a_{31} + a_{13}a_{21}a_{32} -$$

$$-a_{11}a_{23}a_{32} - a_{12}a_{21}a_{33} - a_{13}a_{22}a_{31}.$$

To memorise (2.6) we can use demonstrative diagrams. We replace the elements of the determinant by dots and join the elements which make one term with lines in the right hand side of (2.6). We do have a separate diagram for terms of each parity. This gives us the following diagrams: −

Example 3. We shall calculate the determinant of a third order matrix using (2.6)

$$\begin{vmatrix} 1 & 2 & 3 \\ 2 & -1 & 0 \\ 3 & -2 & 1 \end{vmatrix} = (1\cdot-1\cdot1)+(2\cdot0\cdot3)+(3\cdot2\cdot-2)-$$

$$-(3\cdot-1\cdot3)-(1\cdot0\cdot-2)-(2\cdot2\cdot1)=-8.$$

We can easily evaluate the number of terms needed to evaluate the nth order determinant. The determinant is equal to the sum of all possible products of elements which take one factor from every row and one factor from every column. The sign + or − is determined by the multiplication of the corresponding

Grassmann units. We calculate this number as follows. The element a_{1i} from the first row can be taken in n different ways. Having fixed the element a_{1i}, the element a_{2j} can be taken in n-1 ways. Having fixed elements a_{1i}, a_{2j} the element from the third row a_{3k} can be taken in n-2 way. This process can be continued until all the rows have been used. All in all there exist

$$n! = n \cdot (n-1) \cdot (n-2) \cdot ... \cdot 2 \cdot 1$$

different ways of taking the product of elements which take one factor from every row and one factor from every column. The number of terms needed to define the determinant of second and third order matrices are $2!=2$ and $3!=6$ respectively.

To define the determinant of a fourth order matrix (2.6) will contain $4! = 24$ terms. Therefore it is inconvenient to use this formula. So to calculate the determinant we can use different transformations by making use of the following properties of the determinants.

§3. Properties of Determinants

We shall look at the basic properties of the determinant, which come directly from the equality (2.3). As we shall see in a later chapter, the following properties belong to both rows and columns of matrices. However, in this chapter we shall consider only rows.

1.   If all the elements of a row of a matrix are equal to zero, then the determinant of the matrix is equal to zero.

Proof. We see that if all the elements $a_{ij} = 0$, $j = 1,...,n$ then $X_i = 0$. Consequently,

$$X_1...X_n = 0 \Leftrightarrow 0 = De_1...e_n \Rightarrow D = 0.$$

*2.   After re-arranging any two rows the determinant **has the same magnitude** but changes its sign.*

Proof. If we re-arrange two adjacent rows, then by skew symmetry we have

$$X_i X_j = -X_j X_i$$

Hence the determinant changes its sign. Rearranging two rows i and j is equivalent to rearranging two factors X_j and X_i in the product $X_1 X_2...X_n$. This rearrangement can be done by rearranging $2(i - j) - 1$ adjacent factors. If an odd number of rearrangements is performed the sign of the product changes. Consequently the value of D in (2.3) changes its sign. However, the magnitude remains the same.

3.   If any two rows of a matrix are the same, then the determinant is equal to zero.

Proof. If the product $X_1X_2...X_n$ contains two or more similar factors, it will be equal to zero. The property described above is a consequence of this. We can also prove this with the previous properties. Consider rearranging two identical rows of a matrix. The determinant will change its sign, but since the two rows are identical the determinant is unchanged. $D = -D \Rightarrow D = 0$

4.   If all the elements of the row are multiplied by a scalar λ, then we have the equality

$$X_1...(\lambda X_i)...X_n = \lambda D e_1...e_n.$$

Proof. Since X_i is an element of a real vector space, scalar multiplication is associative and commutative. The above property follows trivially. We can state this in another way; the common factor of the elements of the row is a factor of the determinant.

5.   *If all the elements of one row are propor-*
tional to the elements of another row, then the deter-
minant is equal to zero.

This property is a consequence of properties 3 and 4.

6.   *If one of the rows of the determinant is*
linearly expressed via the other rows then the deter-
minant is equal to zero.

Proof. For example, let the first row of the determinant be a linear combination of the other ones, i.e.

$$X_1 = \lambda_2 X_2 + ... + \lambda_n X_n.$$

For the matrix elements we have the following equality:

$$A_{1s} = \lambda_2 a_{2s} + ... + \lambda_n a_{ns}, s = 1,...,n.$$

Upon substitution of X_1 into (2.3) we see that

$$X_1 X_2 ... X_n = \lambda_2 X_2 X_2 ... X_n + ... + \lambda_n X_n X_2 ... X_n = 0$$
$$\Rightarrow D = 0.$$

Each term contains the product of the linear forms with repeated factors, thus each term is equal to zero.

7.   *The value of the determinant does not change if a multiple of one row of the matrix is added to another.*

Proof. Suppose that   the sth row is added to the rth row λ times. Then we have the equality

$$X_1...\left(X_r + \lambda X_s\right)...X_s...X_n = X_1...X_n + \lambda X_1...X_s...X_s...X_n =$$
$$= X_1...X_n = D.$$

This proves the property, which can be generalised in the following way. If a linear combination of the rows is added to a row of the matrix, then the determinant of the matrix is left unchanged.

8. *Let each element of the ith row of the matrix be composed of the sum of two terms. Then the determinant is equal to the sum of the determinants of the two matrices whose rows coincide with the original matrix except for the ith row, and such that the sum of the ith rows of the two matrices equals the ith row of the original matrix. For i=1 this condition has the following formulation*

$$\begin{vmatrix} b_{11}+c_{11} & b_{12}+c_{12} & \dots & b_{1n}+c_{1n} \\ a_{21} & a_{22} & \dots & a_{2n} \\ \dots & \dots & \dots & \dots \\ a_{n1} & a_{n2} & \dots & a_{nn} \end{vmatrix} = \begin{vmatrix} b_{11} & b_{12} & \dots & b_{1n} \\ a_{21} & a_{22} & \dots & a_{2n} \\ \dots & \dots & \dots & \dots \\ a_{n1} & a_{n2} & \dots & a_{nn} \end{vmatrix} + \begin{vmatrix} c_{11} & c_{12} & \dots & c_{1n} \\ a_{21} & a_{22} & \dots & a_{2n} \\ \dots & \dots & \dots & \dots \\ a_{n1} & a_{n2} & \dots & a_{nn} \end{vmatrix}.$$

Proof. The proof of this property comes trivially from the equality

$$X_1 \dots \left(X_i' + X_i'' \right) \dots X_n = X_1 \dots X_i' \dots X_n + X_1 \dots X_i'' \dots X_n.$$

More complex properties will be proved in the following chapters.

§4. Calculation of Determinants

Let all elements of the matrix lying below the diagonal equal zero. The determinant of such a matrix is equal to the product of the diagonal elements, i.e.

$$\begin{vmatrix} a_{11} & a_{12} & a_{13} & \ldots & a_{1n} \\ 0 & a_{22} & a_{23} & \ldots & a_{2n} \\ 0 & 0 & a_{33} & \ldots & a_{3n} \\ \ldots & \ldots & \ldots & \ldots & \ldots \\ 0 & 0 & 0 & \ldots & a_{nn} \end{vmatrix} = a_{11}a_{22}a_{33}\ldots a_{nn}.$$

(4.1)

This is easily proved by multiplying the corresponding linear forms. Using properties 2 and 4, which were discussed in the previous chapter, it is possible to make any determinant take the form shown above.

Example 4. Calculate the determinant

$$\begin{vmatrix} 2 & 1 & 3 \\ 0 & 2 & 3 \\ 4 & 4 & 6 \end{vmatrix} = \begin{vmatrix} 2 & 1 & 3 \\ 0 & 2 & 3 \\ 0 & 2 & 0 \end{vmatrix} = \begin{vmatrix} 2 & 1 & 3 \\ 0 & 2 & 3 \\ 0 & 0 & -3 \end{vmatrix} = -12.$$

The elements in the first row are multiplied by −2 and added to the corresponding elements of the third row. Then the elements of the second row are deducted from the corresponding elements of the third row. As a result we get a matrix whose elements below the diagonal are equal to zero and the value of the determinant is calculated according to (4.1).

§5. Equivalence of Determinants

We shall see how calculation of the determinant with the help of (2.3) coincides with the conventional definition. We take a product of different Grassmann units $e_{s1}, ..., e_{sn}$ that are taken at a random order. The indices $s1, s2, ..., sn$ are a permutation of the integers $1, 2, ..., n$. We shall denote the number of inversions in the sequence $s1, s2, ..., sn$ by S, that is the number of all cases when a bigger number appears before the smaller number. Every inversion of adjacent factors changes the sign of the product.   Starting from $e_1 e_2 ... e_n$ where the number of inversions is zero, each inversion required to arrive at $e_{s1} ... e_{sn}$ changes the sign of the product. This gives the formula

$$e_{s1} ... e_{sn} = \left(-1\right)^S e_1 ... e_n.$$

(5.1)

Example 5. The permutation $4, 2, 1, 3$ has a value S equal to $3+1+0=4$. Thus we have the equality

$$e_4 e_2 e_1 e_3 = e_1 e_2 e_3 e_4.$$

The permutation $s1, s2, ..., sn$ is called even if the number of inversions is even, and it is called odd if the number of inversions is odd. We transform the multiplication of the linear forms X_i as in (2.2)

$$X_1...X_n = \sum_{s_i \in S_n} a_{x_1 s_1}...a_{x_n s_n} e_{s_1}...e_{s_n} = \sum_{s_i \in S_n} (-1)^S a_{x_1 s_1}...a_{x_n s_n} e_1...e_n.$$

Thus we arrive at a well-known formula for the determinants of $n \times n$ matrices

$$D = \det A = \sum_{s_i \in S_n} (-1)^S a_{x_1 s_1}...a_{x_n s_n},$$

(5.2)

where the indices s_i form all permutations of the integers from $1,...,n$. The number S is equal to the number of inversions in the permutation s_i. We can also consider permutations amongst the factors of the product. We introduce the permutation $x_j \in S_n$ and denote the number of inversions of x_j by X. From the properties of linear forms we get the following result.

$$X_1...X_n = (-1)^X X_{x_1}...X_{x_n} = \sum_{s_i \in S_n} (-1)^X a_{x_1 s_1}...a_{x_n s_n} e_{s_1}...e_{s_n}$$

$$= \sum_{s_i \in S_n} (-1)^{X+S} a_{x_1 s_1}...a_{x_n s_n} e_1...e_n.$$

In terms of determinant theory, this result can be expressed as

$$D = \det A = \sum_{s_i \in S_n} (-1)^{X+S} a_{x_1 s_1} \dots a_{x_n s_n},$$

(5.3)

which is the usual formula for calculating determinants. The value of the determinant is equal to the sum of all possible products consisting of the matrix elements taken one by one from each row and each column; they have a positive sign if the permutations of the first and second indices have the same parity, and they have a negative sign if the permutations of the first and second indices have different parity.

For every term in (5.3) one can rearrange the factors so that the second indices will increase. It comes out that

$$D = \sum_{x_j \in S_n} (-1)^X a_{x_1 1} \dots a_{x_n n}.$$

(5.4)

This means that the value of the determinant does not change if we take the row elements as the coefficients of the linear forms. We have

$$Y_s = a_{1s} e_1 + \dots + a_{ns} e_n, \quad s = 1, \dots, n.$$

Multiplication of these linear forms leads to

$$Y_1...Y_n = \sum_{x_j \in S_n} a_{x_1 1}...a_{x_n n} e_{x_1}...e_{x_n} =$$

$$= \sum_{x_j \in S_n} (-1)^X a_{x_1 1}...a_{x_n n} e_1...e_n = De_1...e_n.$$

Thus we come to the most important property of the determinant:

$$\left| a_{ij} \right| = \left| a_{ji} \right| \Leftrightarrow \det A = \det A^T.$$

(5.5)

The operation defined in (5.5) is called the transpose mapping of a determinant. The transpose maps the rows of a determinant to the columns and vice versa. A similar operation is used with matrices. The transpose of a matrix A is denoted by A^T.

9. Taking the transpose of a determinant does not change its value.

This property demonstrates the equality of the rows and columns. An immediate corollary of this property is that properties 1-8 are true for both rows and columns.

§6. Determinant Multiplication

We now consider values D_1, D_2 for determinants $|a_{xs}|^n_1$, $|b_{s1}|^n_1$. We introduce the linear forms

$$X_{is} = \sum_{i=1}^{n} a_{is} Y_s, \quad Y_s = \sum_{r=1}^{n} b_{sr} e_r.$$

$$(6.1)$$

Linear forms Y_s can be considered as generalised Grassmann numbers which satisfy the following condition

$$Y_s Y_q = -Y_q Y_s, \qquad s, q = 1, ..., n.$$

So eliminating Y_s we find the following equality

$$X_1 ... X_n = D_1 Y_1 ... Y_n = D_1 D_2 e_1 ... e_n$$

$$(6.2)$$

Expressing X_i directly in terms of e_1, e_2, ... e_n, we get

$$X_i = \sum_{s=1}^{n} a_{is} \sum_{r=1}^{n} b_{sr} e_r = \sum_{r=1}^{n} \left(\sum_{s=1}^{n} a_{is} b_{sr} \right) e_r = \sum_{r=1}^{n} c_{ir} e_r,$$

where

$$c_{ir} = \sum_{s=1}^{n} a_{is} b_{sr}.$$

$$(6.3)$$

We define the determinant $|c_{xl}|^n_l$ as D and find

$$X_1...X_n = De_1...e_n.$$

$$(6.4)$$

Comparing equalities (6.2) and (6.4) leads to the following property of the determinant.

   10.   *Consider matrices A,B and their multiplication-matrix C.*

$$A = \begin{pmatrix} a_{11} & a_{12} & ... & a_{1n} \\ a_{21} & a_{22} & ... & a_{2n} \\ ... & ... & ... & ... \\ a_{n1} & a_{2n} & ... & a_{nn} \end{pmatrix},$$

$$B = \begin{pmatrix} b_{11} & b_{12} & ... & b_{1n} \\ b_{21} & b_{22} & ... & b_{2n} \\ ... & ... & ... & ... \\ b_{n1} & b_{2n} & ... & b_{nn} \end{pmatrix},$$

$$C = \begin{pmatrix} c_{11} & c_{12} & ... & c_{1n} \\ c_{21} & c_{22} & ... & c_{2n} \\ ... & ... & ... & ... \\ c_{n1} & c_{2n} & ... & c_{nn} \end{pmatrix} = AB,$$

$$c_{ir} = a_{i1}b_{1r} + ... + a_{in}b_{nr}.$$

The determinant of the product of these matrices is equal to the product of the determinants, i.e.

$$\det C = \det A \cdot \det B.$$

$$(6.5)$$

The determinant of matrix C is called the product of the determinants of matrices A and B. Transposing matrices A and B, i.e. rearranging the rows into columns we find another three kinds of determinant multiplication.

Example 6. Verify property 10 with the following two 2×2 matrices.

$$A = \begin{pmatrix} 1 & 2 \\ 3 & 4 \end{pmatrix}, \ \det A = -2,$$

$$B = \begin{pmatrix} 2 & 1 \\ 5 & 3 \end{pmatrix}, \ \det B = 1.$$

We calculate the determinant of the matrix multiplication

$$\det AB = \begin{vmatrix} 12 & 7 \\ 26 & 15 \end{vmatrix} = -2,$$

$$\det AB^T = \begin{vmatrix} 4 & 11 \\ 10 & 27 \end{vmatrix} = -2,$$

$$\det A^T B = \begin{vmatrix} 17 & 10 \\ 24 & 14 \end{vmatrix} = -2,$$

$$\det A^T B^T = \begin{vmatrix} 5 & 14 \\ 8 & 22 \end{vmatrix} = -2.$$

Note 2. If A is a regular matrix, i.e. $\det A \neq 0$, then $A \cdot A^{-1} = I$, where I is the identity matrix. All elements along the diagonal of the identity matrix are equal to one and the rest of the elements are equal to zero. From property 10 we see that

$$\det A \cdot \det A^{-1} = \det I = 1.$$

Therefore it follows that

$$\det A^{-1} = \left(\det A \right)^{-1}.$$

$$(6.6)$$

§7. Expansion of Determinants in Terms of Rows and Columns

Consider the nth order determinant

$$D = \begin{vmatrix} a_{11} & a_{12} & ... & a_{1n} \\ a_{21} & a_{22} & ... & a_{2n} \\ ... & ... & ... & ... \\ a_{n1} & a_{2n} & ... & a_{nn} \end{vmatrix}.$$

We remove one row and one column from the determinant. The row and column intersect at the element a_{is}. The result is an $(n-1)^{th}$ order determinant called the minor and is denoted by M_{is}.

$$M_{is} = \begin{vmatrix} a_{11} & ... & a_{1(s-1)} & a_{1(s+1)} & ... & a_{1n} \\ ... & ... & ... & ... & ... & ... \\ a_{(i-1)1} & ... & a_{(i-1)(s-1)} & a_{(i-1)(s+1)} & ... & a_{(i-1)n} \\ a_{(i+1)1} & ... & a_{(i+1)(s-1)} & a_{(i+1)(s+1)} & ... & a_{(i+1)n} \\ ... & ... & ... & ... & ... & ... \\ a_{n1} & ... & a_{n(s-1)} & a_{n(s+1)} & ... & a_{nn} \end{vmatrix}.$$

We define the term

$$A_{is} = (-1)^{i+s} M_{is}, \qquad i,s = 1,...,n,$$

$$(7.1)$$

as the algebraic addition to element a_{is}.

Example 7. We shall calculate some minors and algebraic additions to the determinant elements:

$$D = \begin{vmatrix} 1 & 2 & 5 \\ -1 & 2 & 1 \\ 2 & -2 & 4 \end{vmatrix}$$

Using the previous definitions we find

$$M_{11} = \begin{vmatrix} 2 & 1 \\ -2 & 4 \end{vmatrix} = 10, \qquad M_{12} = \begin{vmatrix} -1 & 1 \\ 2 & 4 \end{vmatrix} = -6,$$

$$M_{13} = \begin{vmatrix} -1 & 2 \\ 2 & -2 \end{vmatrix} = -2,$$

$$A_{11} = (-1)^{1+1} M_{11} = 10, \qquad A_{12} = (-1)^{1+2} M_{12} = 6,$$

$$A_{13} = (-1)^{1+3} M_{13} = -2.$$

We express these last results in terms of linear forms (2.2). The definition of the determinant according to (2.3) is applied to the $(n-1)^{th}$ order determinants and is found by multiplication of $(n-1)$ rows:

$$X_1...X_{i-1}X_{i+1}...X_n = M_{i1}e_2e_3...e_n + ... + M_{in}e_1e_2...e_{n-1}.$$

Consider the product of all linear forms

$$X_1...X_{i-1}X_iX_{i+1}...X_n = (-1)^{(i+1)} X_iX_1...X_{i-1}X_{i+1}...X_n =$$

$$= (-1)^{(i+1)} (a_{i1}e_1 + ... + a_{in}e_n)(M_{i1}e_2e_3...e_n + ... + M_{in}e_1e_2...e_{n-1}) =$$

$$= ((-1)^{(i+1)} a_{i1}M_{i1} + ... + (-1)^{(i+n)} a_{in}M_{in})e_1...e_n.$$

Making use of (7.1) and (2.3) we find the equality

$$D = a_{i1}A_{i1} + ... + a_{in}A_{in},$$

(7.2)

which expresses the following property of the determinant:

11.   The determinant is equal to the sum of products of all the elements of a row or column with their algebraic additions. Therefore the determinant is a linear homogeneous function of the elements of a row or a column, and the algebraic additions of the elements are the coefficients of this linear function decomposed in terms of the row or column elements, i.e.

$$\frac{\partial D}{\partial a_{is}} = A_{is}, \qquad i, s = 1, ..., n.$$

The right hand side of (7.2) is called the determinant expansion into the elements of the ith row. If we replace the ith row by the qth row in the determinant D, i.e. we set

$$a_{is} = a_{qs}, \qquad s = 1,...,n; q \neq i,$$

then the value of the determinant is equal to zero according to property 4.

From (7.2) we see that

$$a_{q1}A_{i1} + ... + a_{qn}A_{in} = 0, \qquad q \neq i,$$

$$(7.3)$$

which proves the following property of determinants.

12. The sum of the product of any row (column) with the algebraic additions of another row (column) is equal to zero.

Property 11 is often used for calculating the determinants.

Example 8. We shall expand the determinant into the elements of the second row.

$$\begin{vmatrix} 1 & 2 & 3 \\ 0 & 1 & 2 \\ -1 & 4 & 3 \end{vmatrix} = -0\begin{vmatrix} 2 & 3 \\ 4 & 3 \end{vmatrix} + 1\begin{vmatrix} 1 & 3 \\ -1 & 3 \end{vmatrix} - 2\begin{vmatrix} 1 & 2 \\ -1 & 4 \end{vmatrix} = -6.$$

To simplify calculations, the determinant should be expanded into elements of the row of column with the most zeros.

§8. Laplace's Theorem

Laplace's theorem generalises property 11 of the determinants. In this chapter we shall use another definition for minors which is different from the definitions used in the previous chapter. We define the term

$$M_{i_1,\ldots,i_q,s_1,\ldots,s_q},$$

(8.1)

as the minor created by the elements which lie at the crossing of the rows. We take the rows with numbers $i_1,\ldots,i_q$ and columns with numbers $s_1,\ldots,s_q$. Note that this definition allows for the removal of multiple rows and columns from the original determinant. We arrange the rows $i_1,\ldots,i_q$ in increasing order. The multiplication of the corresponding linear forms is

$$X_{i_1}\ldots X_{i_q} = \sum_{s_1,\ldots,s_q} M_{i_1,\ldots,i_q,s_1,\ldots,s_q} e_{s_1}\ldots e_{s_q},$$

(8.2)

where indices $s_1,\ldots,s_q$ are also arranged in increasing order and are chosen from the set of indices $1,2,\ldots,n$. For multiplication of the rest rows with increasing numbers $i_{q+1}\ldots i_n$ we have a similar expression, namely

$$X_{i_{q+1}}...X_n = \sum_{s_{q+1},...,s_n} M_{i_{q+1},...,i_n,s_{q+1},...,s_n} e_{q+1}...e_n.$$

(8.3)

Suppose that indices $s_{q+1},...,s_n$ are arranged in increasing order and are chosen from indices $1,2,...,n$. When we multiply expressions (8.2) and (8.3) the non-zero components are received only if indices $s_1,...,s_q, s_{q+1},...,s_n$ is a permutation of the integers $1,2,...,n$.

Using this result we shall find the following property of the Grassmann units

$$e_{s_1}...e_{s_q} e_{s_{q+1}}...e_{s_n} = (-1)^{(s_1-1)+(s_2-2)+...+(s_q-q)} e_1...e_n.$$

(8.4)

In order to prove this formula we construct the left hand side of the equality from the product $e_1 e_2...e_n$. In order to do this we invert e_{s_1} with s_1-1 factors to bring it to the front of the product. Then we invert e_{s_2} with s_2-2 factors to bring it to the second position in the product, we repeat this process until we have all the elements in the correct positions. Since $s_1 < s_2$ and $s_2 > 1$, this process is well defined. The structure follows the inequalities $s_1 < ... < s_q$,

$s_{q+1} < \ldots < s_n$ and we shall obtain the number of inversions shown in (8.4).

Using a similar formula for the multiplication of the linear forms $X_1, X_2, \ldots, X_n$ we find the equality

$$X_1 \ldots X_n = (-1)^{(i_1 - 1) + \ldots + (i_q - 1)} X_{i_1} \ldots X_{i_q} X_{i_{q+1}} \ldots X_{i_n}.$$

(8.5)

The expression

$$A_{i_1, \ldots, i_q, s_i, \ldots, s_q} = (-1)^{i_1 + \ldots + i_q + s_1 + \ldots + s_q} M_{i_{q+1}, \ldots, i_n, s_{q+1}, \ldots, s_n}$$

(8.6)

is defined as the algebraic addition to minor (8.1). In order to calculate the algebraic addition we take the minor additional to minor (8.1). To do this we shall remove rows $i_1, \ldots, i_q$ and columns $s_1, \ldots, s_q$ from the determinant. The sign of the minor is defined according to (8.6) and is positive if the sum of the numbers of all the crossed out rows and columns is even and is negative if the sum is odd.

Comparing equalities (2.3) and (8.5) we come to the formula

$$D = \sum_{s_1, \ldots, s_q} M_{i_1, \ldots, i_q, s_1, \ldots, s_q} A_{i_1, \ldots, i_q, s_i, \ldots, s_q},$$

(8.7)

which is the well-known Laplace theorem.

Theorem. If we choose q rows (q columns) in a nth order determinant D (where $q<n$). Then the sum of the products of the qth order minors over the chosen rows (columns) with their algebraic additions is equal to the determinant D.

Example 9. We shall calculate the determinant

$$\begin{vmatrix} 1 & 0 & 0 & 2 \\ 0 & 1 & 0 & 3 \\ 1 & 0 & 0 & 1 \\ 2 & 3 & 4 & 1 \end{vmatrix},$$

using Laplace's theorem. We choose the 1st and the 3rd rows and obtain

$$\begin{vmatrix} 1 & 0 & 0 & 2 \\ 0 & 1 & 0 & 3 \\ 1 & 0 & 0 & 1 \\ 2 & 3 & 4 & 1 \end{vmatrix} = (-1)^{1+3+1+4} \begin{vmatrix} 1 & 2 \\ 1 & 1 \end{vmatrix} \cdot \begin{vmatrix} 1 & 0 \\ 3 & 4 \end{vmatrix} = 4.$$

All the terms not shown are equal to zero.

A particular case of Laplace's theorem, when only one row or one column is chosen, gives the formula for the ith row as

$$D = a_{i1}A_{i1} + ... + a_{in}A_{in},$$

which refers to what we found in Chapter 7.

Note 3. Laplace's theorem can be applied for simplifying a determinant of the kind

$$D = \det \begin{pmatrix} A & C \\ 0 & B \end{pmatrix},$$

where A is an $n \times n$ matrix, B is an $m \times m$ matrix, C is an $n \times m$ matrix and 0 is an $m \times n$ matrix with zero elements. Choosing the first n rows we can see that only one of the algebraic additions is different from zero. Consequently, we obtain the formula

$$D = \det A \cdot \det B.$$

Example 10. Rearranging firstly the rows and then the columns of the determinant, we calculate its value as

$$\begin{vmatrix} 1 & 0 & 0 & 2 \\ 0 & 1 & 0 & 3 \\ 1 & 0 & 0 & 1 \\ 2 & 3 & 4 & 1 \end{vmatrix} = - \begin{vmatrix} 0 & 1 & 0 & 3 \\ 2 & 3 & 4 & 1 \\ 1 & 0 & 0 & 2 \\ 1 & 0 & 0 & 1 \end{vmatrix} = - \begin{vmatrix} 1 & 0 & 0 & 3 \\ 3 & 4 & 2 & 1 \\ 0 & 0 & 1 & 2 \\ 0 & 0 & 1 & 1 \end{vmatrix} = - \begin{vmatrix} 1 & 0 \\ 3 & 4 \end{vmatrix} \begin{vmatrix} 1 & 2 \\ 1 & 1 \end{vmatrix} = 4.$$

Note 4. The problem of estimating determinants often appears, for example when estimating electromechanical schemes. In some cases many of the elements are equal to zero, allowing us to use Grassmann numbers directly for the estimation of determinants.

Example 11. We shall estimate the following fourth order determinant.

$$D = \begin{vmatrix} 1 & 0 & 2 & 0 \\ 0 & 1 & 0 & 2 \\ 0 & 3 & 1 & 0 \\ 2 & 0 & 0 & 3 \end{vmatrix}.$$

We introduce the expressions

$$X_1 = e_1 + 2e_3, \quad X_2 = e_2 + 2e_4, \quad X_3 = 3e_2 + e_3,$$
$$X_4 = 2e_1 + 3e_4$$

to find the product

$$X_1 X_2 X_3 X_4 = (X_1 X_2)(X_3 X_4) = (3e_1e_4 - 4e_1e_3 + 6e_3e_4)(e_2e_3 - 6e_2e_2 - 2e_3e_4) =$$
$$= 3e_1e_4e_2e_3 + 24e_1e_3e_2e_4 = -21e_1e_2e_3e_4, \quad D = -21.$$

To conclude we note that these properties also hold for determinants with complex elements. In the next chapter we will explain some ways to solve systems of linear algebraic equations using determinant theory.

§9. Solutions of Linear Systems of Equations

There is a solution of the system of n linear algebraic equations with n unknown $x_1, x_2, \ldots, x_n$

$$a_{11}x_1 + \ldots + a_{1n}x_n = b_1,$$

$$a_{21}x_1 + \ldots + a_{2n}x_n = b_2,$$

$$\ldots\ldots\ldots\ldots\ldots\ldots\ldots\ldots\ldots\ldots$$

$$a_{n1}x_1 + \ldots + a_{nn}x_n = b_n.$$

$$(9.1)$$

We introduce the following definitions. The nth order determinant

$$D = \det A = \begin{vmatrix} a_{11} & a_{12} & \ldots & a_{1n} \\ a_{21} & a_{22} & \ldots & a_{2n} \\ \ldots & \ldots & \ldots & \ldots \\ a_{n1} & a_{n2} & \ldots & a_{nn} \end{vmatrix}$$

$$(9.2)$$

is called the determinant of system (9.1). We shall denote by D_s the determinant whose s^{th} column is replaced by the right hand side of system (9.1). Now we shall prove Kramer's theorem.

Theorem. If determinant D of system (9.1) is non-zero, then the system has an unique solution

$$x_s = \frac{D_s}{D}, \qquad s = 1,...,n.$$

$$(9.3)$$

Proof. We multiply each equation of system (9.1) by algebraic addition A_{is} of the sth column of determinant D. By summing the new equations we get

$$\left(a_{11}A_{1s} + ... + a_{n1}A_{ns}\right)x_1 + \left(a_{12}A_{2s} + ... + a_{n2}A_{ns}\right)x_2 + ... +$$
$$+ \left(a_{31}A_{1s} + ... + a_{n3}A_{ns}\right)x_n = b_1A_{1s} + ... + b_nA_{ns}.$$

According to properties 11 and 12 these equations can be written in the form

$$Dx_s = D_s, \quad s = 1,...,n,$$

$$(9.4)$$

from which we can receive solution (9.3). In order to prove the equivalence of the system of equations (9.4) to system (9.1) we write (9.4) as

$$Dx_s = b_1A_{1s} + ... + b_nA_{ns}, \quad s = 1,...,n.$$

We then multiply each side of the sth equation by a_{is} and sum the results, thereby gaining the equations

$$D(a_{i1}x_1 + ... + a_{in}x_n) = b_1(a_{i1}A_{11} + ... + a_{i1}A_{11}) + ... +$$
$$+ b_n(a_{in}A_{n1} + ... + a_{in}A_{nn}), \quad i = 1,...,n,$$

which according to properties 11 and 12 can be written as

$$D(a_{i1}x_1 + ... + a_{in}x_n) = b_i D, \quad i = 1,...,n.$$

If $D \neq 0$ this system of equations coincides with the initial system (9.1).

Example 12. We find the solution of the following system of two liner algebraic equations

$$2x_1 + 3x_2 = 1,$$
$$x_1 + 4x_2 = 3.$$

Using (9.3) we find the solution as

$$x_1 = \frac{\begin{vmatrix} 1 & 3 \\ 3 & 4 \end{vmatrix}}{\begin{vmatrix} 2 & 3 \\ 1 & 4 \end{vmatrix}} = \frac{-5}{5} = -1, \qquad x_2 = \frac{\begin{vmatrix} 2 & 1 \\ 1 & 3 \end{vmatrix}}{\begin{vmatrix} 2 & 3 \\ 1 & 4 \end{vmatrix}} = \frac{5}{5} = 1.$$

Note 4. Kramer's formula (9.8) can be applied to find a solution of system (9.1) for small values of n. For systems of higher order, application of these formulas is not productive since they are expensive. Usually Gauss's method is used for larger systems.

§10. Interpolation Polynomial Construction

We now consider the problem of finding an interpolation polynomial, i.e the polynomial

$$P(x) = a_0 + a_1 x + \ldots + a_n x^n,$$

(10.1)

which for the points $x_0, x_1, \ldots, x_n$ will take the values $y_0, y_1, \ldots y_n$. In order to define the coefficients of the polynomial we construct an $(n+1)^{th}$ order system of linear algebraic equations

$$P(x_i) = a_0 + a_1 x_i + \ldots + a_n x_i^n = y_i, \quad i = 1, \ldots, n.$$

(10.2)

Estimating the determinant of this system

$$D = \begin{vmatrix} 1 & x_0 & \ldots & x_0^n \\ 1 & x_1 & \ldots & x_1^n \\ \ldots & \ldots & \ldots & \ldots \\ 1 & x_n & \ldots & x_n^n \end{vmatrix}$$

(10.3)

is called the Vandermonde determinant. If we subtract the first row from the rest of the rows and expand it into the elements of the first column, we get the following determinant

$$D_{n+1} = \begin{vmatrix} x_1 - x_0 & \cdots & x_1^n - x_0^n \\ \cdots & \cdots & \cdots \\ x_n - x_0 & \cdots & x_n^n - x_0^n \end{vmatrix}.$$

We then take the common factor $x_i - x_0$ from each row to get

$$D_{n+1} = (x_1 - x_0)\ldots(x_n - x_0)\begin{vmatrix} 1 & x_1 + x_0 & x_1^{n-1} + x_1^{n-2}x_0 + \ldots + x_0^{n-1} \\ \cdots & \cdots & \cdots \\ 1 & x_n + x_0 & x_n^{n-1} + x_n^{n-2}x_0 + \ldots + x_0^{n-1} \end{vmatrix}.$$

By subtracting x_0^i from the ith column, the determinant is found to be of the same type as (10.3), but of a lower order

$$D_{n+1} = (x_1 - x_0)\ldots(x_n - x_0)\begin{vmatrix} 1 & \cdots & x_1^{n-1} \\ \cdots & \cdots & \cdots \\ 1 & \cdots & x_n^{n-1} \end{vmatrix}.$$

$$(10.4)$$

Repeating this reduction we finally come to the formula

$$D_{n+1} = \prod_{0 \le s < i \le n} (x_i - x_s).$$

$$(10.5)$$

From (10.5) we can conclude that the determinant of the system is non-zero if $x_i \ne x_s, \ \forall i > s$.

Our final conclusion is that the interpolation polynomial (10.1) can only be constructed if the number of different points – the interpolation variables x_0, x_1, ...,x_n - is equal to the number of unknown coefficients a_0, a_1,..., a_n of the polynomial.

The following two examples were created by **Heide Gieber (Austria, Vienna University of Technology)** and **Peter Smith (UK, Queen's University Belfast).**

Example. We use (10.5) to find the value of the following determinant;

$$D = \begin{vmatrix} 1 & 4 & 16 & 64 \\ 1 & 5 & 25 & 125 \\ 1 & 2 & 4 & 8 \\ 1 & 3 & 9 & 27 \end{vmatrix},$$

where $x_0 = 4, x_1 = 5, x_2 = 2, x_3 = 3$:

$$D = (5-4)(2-4)(3-4)(2-5)(3-5)(3-2) = 12.$$

This answer is easily verified by subtracting the third row from each of the others, and then expanding the determinant into the first column.

Example. The following is a demonstration of the practical uses of determinant theory. An electronics experiment was carried out, measuring the uncertainty in voltage for a selection of electronic devices.

The uncertainty as a function of frequency was measured for each device. Second order polynomial functions were fitted to the graphs, and these were used to calculate the determinants of the systems.

Measurement errors at given frequencies 60, 80 and 100MHz for 5 different devices are:

1) PPITE 0.5V: $(y_0 \quad y_1 \quad y_2) = (-1.42 \quad -2.10 \quad -2.35)$

2) EPITE 0.5V: $(y_0 \quad y_1 \quad y_2) = (-1.37 \quad -1.92 \quad -2.15)$

3) DTPT-6 EPITE 0.5V:

$(y_0 \quad y_1 \quad y_2) = (-0.62 \quad -0.57 \quad -0.81)$

4) EPITE 1V: $(y_0 \quad y_1 \quad y_2) = (-0.51 \quad -0.45 \quad -0.42)$

5) PPITE 1V: $(y_0 \quad y_1 \quad y_2) = (-0.30 \quad -0.19 \quad -0.15)$

To estimate measurement errors at other frequencies (10, 20 and 30MHz) the interpolation polynomial was calculated at those frequency values.

We obtain the interpolation polynomial of 2nd order (this order is sufficient for the purpose of this demonstration):

$$P(x) = a_0 + a_1 x + a_2 x^2$$

The given observations at three different frequencies are:

$$x_0 = 60 \qquad\qquad P(x_0) = y_0$$
$$x_1 = 80 \qquad\qquad P(x_1) = y_1$$
$$x_2 = 100 \qquad\qquad P(x_2) = y_2$$

We get a 2nd order system of linear algebraic equations like in (10.2), i.e. the Vandermonde-determinant:

$$D = \begin{vmatrix} 1 & x_0 & x_0^2 \\ 1 & x_1 & x_1^2 \\ 1 & x_2 & x_2^2 \end{vmatrix}.$$

The coefficients of the interpolation polynomial can be calculated as follows:

$$a_0 = D_0/D$$
$$a_1 = D_1/D$$
$$a_2 = D_2/D,$$

where

$$D_0 = \begin{vmatrix} y_0 & x_0 & x_0^2 \\ y_1 & x_1 & x_1^2 \\ y_2 & x_2 & x_2^2 \end{vmatrix}, \quad D_1 = \begin{vmatrix} 1 & y_0 & x_0^2 \\ 1 & y_1 & x_1^2 \\ 1 & y_2 & x_2^2 \end{vmatrix},$$

$$D_2 = \begin{vmatrix} 1 & x_0 & y_0 \\ 1 & x_1 & y_1 \\ 1 & x_2 & y_2 \end{vmatrix}.$$

Once we found the interpolation polynomial, we can find values for other frequencies using extrapolation (i.e. calculation of $P(x_i)$).

1) PPITE 0.5V:

$$(y_0 \quad y_1 \quad y_2) = (-1.42 \quad -2.10 \quad -2.35)$$

$D = 16'000 \qquad P(x) = 3.2 - 0.10925 \cdot x + 0.0005375 \cdot x^2$

$D_0 = 51'200 \qquad P(10) = 2.16125$

$D_1 = -1'748 \qquad P(20) = 1.23$

$D_2 = 8.6 \qquad\quad P(30) = 0.40625$

2) EPITE 0.5V:

$$(y_0 \quad y_1 \quad y_2) = (-1.37 \quad -1.92 \quad -2.15)$$

$D = 16'000 \qquad\qquad P(x) = 2.2 - 0.0835 \cdot x + 0.0004 \cdot x^2$

$D_0 = 35'200 \qquad\qquad P(10) = 1.405$

$D_1 = -1'336 \qquad\qquad P(20) = 0.690$

$D_2 = 6.4 \qquad\qquad\quad P(30) = 0.055$

3) DTPT-6 EPITE 0.5V:

$$(y_0 \quad y_1 \quad y_2) = (-0.62 \quad -0.57 \quad -0.81)$$

$D = 16'000$

$D_0 = -40'160$

$D_1 = 852$

$D_2 = -5.8$

$P(x) = -2.51 + 0.05325 \cdot x - 0.0003625 \cdot x^2$

$P(10) = -2.01375$

$P(20) = -1.59$

$P(30) = -1.23875$

4) EPITE 1V:

$$(y_0 \quad y_1 \quad y_2) = (-0.51 \quad -0.45 \quad -0.42)$$

$D = 16'000$

$D_0 = -13'920$

$D_1 = 132$

$D_2 = -0.6$

$$P(x) = -0.87 + 0.00825 \cdot x - 0.0000375 \cdot x^2$$

$$P(10) = -0.79125$$

$$P(20) = -0.72$$

$$P(30) = -0.65625$$

5) PPITE 1V:

$$(y_0 \quad y_1 \quad y_2) = (-0.30 \quad -0.19 \quad -0.15)$$

$$D = 16'000$$
$$D_0 = -16'800$$
$$D_1 = 284$$
$$D_2 = -1.4$$

$$P(x) = -1.05 + 0.01775 \cdot x - 0.0000875 \cdot x^2$$
$$P(10) = -0.88125$$
$$P(20) = -0.73$$
$$P(30) = -0.59625$$

Below are the functions of uncertainty in voltage as a function of frequency for each of the five devices.

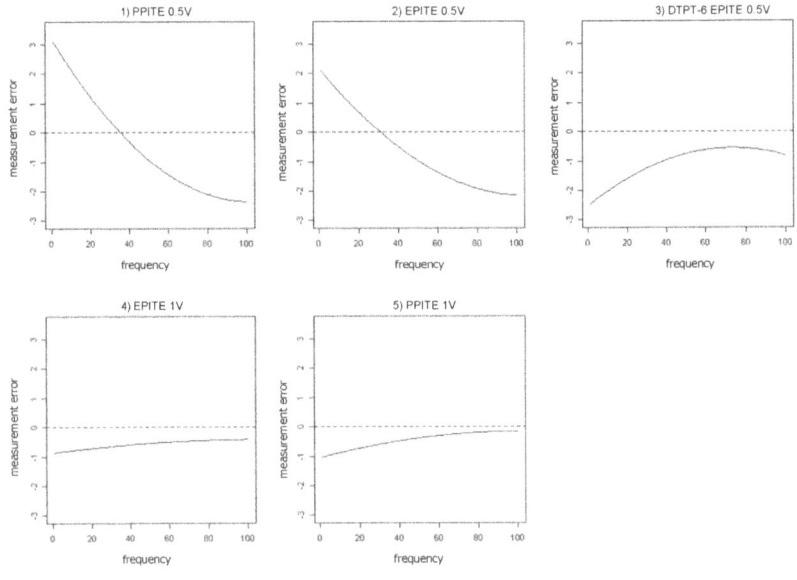

§11. Solutions of Infinite Linear Systems of Equations

The next task shall be to explain a new method of dealing with solutions to linear systems of algebraic equations in infinite dimensions.

The essence of this method is that instead of seeking the solution itself we shall find some linear combination of the unknown units with stated coefficients. Let H be some Hilbert space with scalar product $\langle \varphi, \psi \rangle$ and let A be a linear operator in H. In practice we often come across the problem of finding the solution of an equation of the form

$$A\phi = f,$$

$$(11.1)$$

where φ is an unknown element and $f \in H$. We can find an approximate solution to equation (11.1) by **_Galerkin's_** method as a linear combination of the linearly independent elements $\varphi_1, \varphi_2, \dots.$

$$\overline{\varphi}_n = x_1 \varphi_1 + \dots + x_n \varphi_n$$

$$(11.2)$$

Unknown coefficients $x_1, x_2, ..., x_n$ are found from the orthogonal conditions

$$A\overline{\varphi}_n - f = \delta$$

on elements $\varphi_1, ..., \varphi_n$ or any other linearly independent elements $\psi_1, ..., \psi_n$. We come to the following system of the linear algebraic equations:

$$x_1 \langle \phi_i, A\phi_1 \rangle + ... + x_n \langle \phi_i, A\phi_n \rangle = \langle \phi_i, f \rangle,$$
$$i = 1, ..., n.$$

Analogous to our previous systems, we set

$$a_{is} = \langle \varphi_i, A\varphi_s \rangle, \quad b_i = \langle \varphi_i, f \rangle, \quad i, s = 1, .., n,$$

so that the system can be written in the form

$$a_{i1}x_1 + ... + a_{in}x_n = b_i.$$

$$(11.3)$$

Here, n can be increased to increase the accuracy of the approximation. In general, the determinant of system (11.3) is non-zero. We introduce the following notation – determinants $\alpha_i, \beta_i, \gamma_i$

$$\alpha_1 = a_{11}, \quad \alpha_2 = \begin{vmatrix} a_{11} & a_{12} \\ a_{21} & a_{22} \end{vmatrix}, ..., \alpha_n = \begin{vmatrix} a_{11} & a_{12} & ... & a_{1n} \\ a_{21} & a_{22} & ... & a_{2n} \\ ... & ... & ... & ... \\ a_{n1} & a_{n2} & ... & a_{nn} \end{vmatrix}.$$

$$(11.4)$$

The determinants β_i are derived from determinants α_i by replacing the last column with the terms on the right hand side of (11.3) to get

$$\beta_1 = b_1, \; \beta_2 = \begin{vmatrix} a_{11} & b_1 \\ a_{21} & b_2 \end{vmatrix}, \ldots, \beta_n = \begin{vmatrix} a_{11} & \cdots & a_{1(n-1)} & b_1 \\ a_{21} & \cdots & a_{2(n-1)} & b_2 \\ \cdots & \cdots & \cdots & \cdots \\ a_{n1} & \cdots & a_{n(n-1)} & b_n \end{vmatrix}.$$

(11.5)

The determinants γ_i are derived from determinants α_i by replacing the last row by the row with elements $\varphi_1, \ldots, \varphi_n$ to get

$$\gamma_1 = \varphi_1, \; \gamma_2 = \begin{vmatrix} a_{11} & a_{12} \\ \varphi_1 & \varphi_2 \end{vmatrix}, \ldots, \gamma_n = \begin{vmatrix} a_{11} & a_{12} & \cdots & a_{1n} \\ \cdots & \cdots & \cdots & \cdots \\ a_{(n-1)1} & a_{(n-1)2} & \cdots & a_{(n-1)n} \\ \varphi_1 & \varphi_2 & \cdots & \varphi_n \end{vmatrix}.$$

(11.6)

The process used to solve system (11.3) is based on the following theorem.

Theorem. If the main minors of the determinant of system (11.3) are non-zero, then the solution $x_1, x_2, \ldots, x_n$ of system (11.3) satisfies the equation

$$\sum_{i=1}^{n} x_i \phi_i = \sum_{i=1}^{n} \frac{\beta_i \delta_i}{\alpha_i \alpha_{i-1}}.$$

(11.7)

Proof. The proof is by induction. For $n=1$, we have the equality

$$x_1 \phi_1 = \frac{b_1 \phi_1}{a_{11}}.$$

The validity of the second one is apparent. Let equality (11.7) be true for the system of order $n+1$. We shall prove that it is true for the system of order n. Using Kramer's formula (9.3) we derive value x_n from system (11.3) as

$$x_n = \frac{\beta_n}{\alpha_n}.$$

The system of linear algebraic equations (11.3) can be written as the system of order $n-1$

$$\sum_{s=1}^{n-1} a_{is} x_s = b_i - a_{in} x_n, \qquad i = 1,...,n.$$

Assuming the value x_n is known we shall find solutions x_s as the sum of the solutions of the linear algebraic equation systems

$$\sum_{s=1}^{n-1} a_{is} x_{s1} = b_i, \qquad \sum_{s=1}^{n-1} a_{is} x_{s2} = -a_{in} x_n,$$

$$x_s = x_{s1} + x_{s2}, \qquad i = 1,...,n.$$

Solving the second system of equations using Kramer's Rule we find the equalities

$$x_{s2} = -\frac{x_n}{\alpha_{n-1}} \begin{vmatrix} a_{11} & \cdots & a_{1(s-1)} & a_{1n} & a_{1(s+1)} & \cdots & a_{1(n-1)} \\ \cdots & \cdots & \cdots & \cdots & \cdots & \cdots & \cdots \\ a_{n1} & \cdots & a_{n(s-1)} & a_{(n-1)n} & a_{(n-1)(s+1)} & \cdots & a_{(n-1)(n-1)} \end{vmatrix} =$$

$$= \frac{\beta_n}{\alpha_{n-1}\alpha_n} \begin{vmatrix} a_{11} & \cdots & a_{1,s-1} & a_{1s} & a_{1,s+1} & \cdots & a_{1,n-1} \\ \cdots & \cdots & \cdots & \cdots & \cdots & \cdots & \cdots \\ a_{n-1,1} & \cdots & a_{n-1,s-1} & a_{n-1,s} & a_{n-1,s+1} & \cdots & a_{n-1,n-1} \\ 0 & \cdots & 0 & 1 & 0 & \cdots & 0 \end{vmatrix}.$$

We now use the identity

$$x_n = \frac{\beta_n}{\alpha_n \alpha_{n-1}} \begin{vmatrix} a_{11} & a_{12} & \cdots & a_{1,n-1} & a_{1n} \\ \cdots & \cdots & \cdots & \cdots & \cdots \\ a_{n-1,1} & a_{n-1,2} & \cdots & a_{n-1,n-1} & a_{n-1,n} \\ 0 & 0 & \cdots & 0 & 1 \end{vmatrix}.$$

Multiplying the previous equalities for x_{s2} by φ_2 and adding them we find the relation

$$x_{12}\phi_1 + \ldots + x_{n-1,2}\phi_{n-1} + x_n\phi_n = \frac{\beta_n}{\alpha_n \alpha_{n-1}} \gamma_n.$$

Suppose we have the condition

$$x_{11}\phi_1 + x_{21}\phi_2 + \ldots + x_{n-1,1}\phi_{n-1} = \sum_{i=1}^{n-1} \frac{\beta_i}{\alpha_i \alpha_{i-1}} \gamma_i.$$

Summing the two relations above leads to (11.7), which proves the theorem.

Using the theorem we can define the notion for the approximate solution $\overline{\varphi}_n$ to be the *n-th* approximation.

$$\overline{\phi}_n = \frac{\beta_1\gamma_1}{\alpha_1\alpha_0} + ... + \frac{\beta_n\gamma_n}{\alpha_n\alpha_{n-1}}.$$

As $n \to \infty$, suppose that the approximate solution tends to the actual solution and then we find a series solution for the equation (11.1).

$$\phi = \frac{\beta_1\gamma_1}{\alpha_1\alpha_0} + ... + \frac{\beta_n\gamma_n}{\alpha_n\alpha_{n-1}} + ...$$

(11.8)

The sum (11.8) truncated at order n gives an approximate solution of equation (11.1). If it is possible to approximate the other terms in the series we shall find the estimation of the accuracy of the approximate solution.

Example 13. We shall find the periodic solution of the second order linear differential equation

$$\frac{d^2z(t)}{dt^2} + 2\cos(t)z(t) = -\sin t.$$

We shall find the odd Fourier series for $z(t)$, which is of the form

$$z(t) = x_1 \sin(t) + ... + x_n \sin(nt) + ...$$

Substituting into the differential equation and equating coefficients of *sin(kt)* we get the following infinite linear system of equations

$$
\begin{aligned}
x_1 + x_2 && = 1 \\
x_1 + 4x_2 + x_3 && = 0 \\
x_2 + 9x_3 + x_4 && = 0 \\
x_3 + 16x_4 + x_5 && = 0
\end{aligned}
$$

....

Replacing elements φ_s by functions *sin(st)* in determinants γ_i gives an estimation for the determinants as

$$
|\alpha_i| \ge \frac{1}{2}(i-1)!(i+1)!, \quad |\beta_i| = 1, \qquad |\gamma_i| \le (i!)^2 .
$$

If instead of solving the infinite system we only solve four of the written equations at $x_s = 0$, the error in finding the periodic solution $z(t)$ does not exceed the value

$$
\varepsilon = \sum_{i=3}^{\infty} \left| \frac{\beta_i \gamma_i}{\alpha_i \alpha_{i-1}} \right| \approx 0.02.
$$

Finally we find the approximate periodic solution

$$z(t) = \frac{556}{413}\sin t - \frac{143}{413}\sin 2t + \frac{16}{413}\sin 3t - \frac{1}{413}\sin 4t \pm 0.02.$$

We note that x_1 is accurate to within 10^{-7}. This can be seen by estimating the coefficient at φ_1. We apply the method we have developed to find the smallest root of the equation

$$0 = a_0 + a_1 x + \dots .$$

Multiplying this equation by $x, x^2, x^3, \dots$ and defining $x_i = x^i$, we shall come to the infinite linear system of equations

$$a_1 x_1 + a_2 x_2 + a_3 x_3 + a_4 x_4 + \dots = 0,$$
$$a_0 x_1 + a_1 x_2 + a_{12} x_3 + a_3 x_4 + \dots = 0,$$
$$a_0 x_2 + a_2 x_3 + a_3 x_4 + \dots = 0,$$
$$a_0 x_3 + a_1 x_4 + \dots = 0.$$

From (11.7) we find a value for x_1 by supposing $\varphi_1 = 1$, $\varphi_2 = 0$, $\varphi_3 = 0; \dots$

$$x = \frac{a_0}{a_1} - \frac{a_0^2 a_2}{a_1 \cdot \begin{vmatrix} a_1 & a_2 \\ a_0 & a_1 \end{vmatrix}} - \frac{a_0^3 \cdot \begin{vmatrix} a_2 & a_3 \\ a_1 & a_2 \end{vmatrix}}{\begin{vmatrix} a_1 & a_2 \\ a_0 & a_1 \end{vmatrix} \cdot \begin{vmatrix} a_1 & a_2 & a_3 \\ a_0 & a_1 & a_2 \\ 0 & a_0 & a_1 \end{vmatrix}} - \frac{a_0^4 \cdot \begin{vmatrix} a_1 & a_2 & a_3 \\ a_0 & a_1 & a_2 \\ 0 & a_0 & a_1 \end{vmatrix}}{\begin{vmatrix} a_1 & a_2 & a_3 \\ a_0 & a_1 & a_2 \\ 0 & a_0 & a_1 \end{vmatrix} \cdot \begin{vmatrix} a_1 & a_2 & a_3 & a_4 \\ a_0 & a_1 & a_2 & a_3 \\ 0 & a_0 & a_1 & a_2 \\ 0 & 0 & a_0 & a_1 \end{vmatrix}} - \dots$$

We shall come to the known Whittaker's formula for the non-linear by modulo root of the equation.

Note 5. When we tried to solve the infinite system of linear algebraic equations by Kramer's rule (9.3) we encountered some difficulties connected with the convergence of infinite determinants. In works by Helga Font Koch it was proved that the infinite system of the form

$$x_i + \sum_{s=1}^{\infty} a_{is} x_s = b_i, \qquad i = 1, 2, \ldots,$$

can be solved by Kramer's rule. If the conditions

$$\sum_{s,i=1}^{\infty} \left(a_{is}\right)^2 < \infty, \qquad \sum_{i=1}^{\infty} \left(b_i\right)^2 < \infty$$

are satisfied, then infinite determinants converge and the solution satisfies

$$\sum_{i=1}^{\infty} \left(x_i\right)^2 < \infty.$$

§12. General Linear Systems of Equations

We begin this chapter by defining the rank of a matrix. Consider a general mxn matrix;

$$A = \begin{pmatrix} a_{11} & \cdots & a_{1n} \\ \cdots & \cdots & \cdots \\ a_{m1} & \cdots & a_{mn} \end{pmatrix}.$$

(12.1)

We remove some rows and columns so that the remaining rows and columns will be equal in number. We then calculate the determinant of the remaining rows and columns. The highest order of such a determinant is equal to the lesser of the numbers m and n.

All determinants we construct in this way are called determinants that are a part of A. Suppose that all the determinants of order q which are a part of the matrix are equal to zero. Then all the determinants of a higher order and which are parts of this matrix are also equal to zero since all these determinants can be expressed by the determinants of order q by Laplace's theorem.

The highest order of the determinant A which is non-zero is called the ***rank*** of matrix A. The rank of the matrix does not change if we rearrange the rows or columns, or if we multiply a row or column by **a non-zero scalar**. This statement also holds if we add a multiple of a row (column) to another row (column). These results can be used to transform a determinant A to another determinant B with the same rank. We can set as many elements as possible to zero in determinant B to simplify the calculation of the rank of the determinant.

Example 14. Calculate the rank of matrix A

$$A = \begin{pmatrix} 1 & 2 & 1 & 5 \\ -3 & 1 & 4 & 6 \\ 2 & 1 & 2 & 10 \end{pmatrix}.$$

If we add 3 times the first row to the second, and 2 times the first row to the third we end up with C, where

$$C = \begin{vmatrix} 1 & 2 & 1 & 5 \\ 0 & 7 & 7 & 21 \\ 0 & 0 & 0 & 0 \end{vmatrix}.$$

Now, subtracting columns of the matrix C to maximise the number of zero elements, we come to matrix B, where

$$B = \begin{pmatrix} 1 & 0 & 0 & 0 \\ 0 & 7 & 0 & 0 \\ 0 & 0 & 0 & 0 \end{pmatrix}.$$

We can easily see that the rank of matrix B is equal to 2. Consequently, the rank of matrix A is equal to 2.

We shall now find the solution of a system of m equations with n unknowns

$$a_{11}x_1 + \ldots + a_{1n}x_n = b_1,$$
$$a_{21}x_1 + \ldots + a_{2n}x_n = b_2,$$
$$\ldots\ldots\ldots\ldots\ldots\ldots\ldots\ldots$$
$$a_{m1}x_1 + \ldots + a_{mn}x_n = b_m.$$

$$(12.2)$$

For solving the system we can use the known theorem by Crocker and Chapel.

Theorem. The system of linear equations (12.2) has a definite solution if and only if the rank of matrix A is equal to the rank of the extended matrix $\overline{A}$ where

$$\bar{A} = \begin{pmatrix} a_{11} & \cdots & a_{1n} & b_1 \\ \cdots & \cdots & \cdots & \cdots \\ a_{n1} & \cdots & a_{nn} & b_n \end{pmatrix}.$$

(12.3)

Proof. By reordering the unknown terms and rearranging the equations it is always possible to transform (12.2) to the form where the highest order non-zero determinant which is part of A is situated in the upper left block of A. This determinant is called the main determinant. We denote the order of the main determinant by q. If $q=m=n$, then the proof of the theorem is trivial. If $q=m<n$, then for every value of $x_{m+1}, \ldots, x_n$, system (12.2) has a solution, and under this condition the rank of A is equal to the rank of B.

Next, we suppose that $q<m$. We multiply column b of matrix $\bar{A}$ by x_s for each s and subtract from the last column. As a result we get matrix B with rank q, where

$$B = \begin{pmatrix} a_{11} & \cdots & a_{1n} & c_1 \\ \cdots & \cdots & \cdots & \cdots \\ a_{m1} & \cdots & a_{mn} & c_m \end{pmatrix}.$$

(12.4)

Let system (12.2) have solution $x_1, x_2, ... x_n$. At these values of x_s in (12.4) the elements of the last row are equal to zero. The rank of matrix B and consequently the rank of $\overline{A}$ is equal to q. Therefore the necessary condition is proved. Let the rank of A and consequently that of B equal q. Since the main determinant is non-zero, at any of the values $x_{q+1}, ... x_n$ it is always possible to choose the values $x_a, ..., x_q$ so that $c_1 = 0, ..., c_q = 0$. Consider a determinant of order $q+1$

$$D_i = \begin{pmatrix} a_{11} & ... & a_{1q} & c_1 \\ ... & ... & ... & ... \\ a_{q1} & ... & a_{qq} & c_q \\ a_{i1} & ... & a_{iq} & c_i \end{pmatrix}, \qquad i = q+1, ..., m .$$

All determinants are equal to zero. Expanding the determinants in terms of the last row gives

$$\begin{vmatrix} a_{11} & ... & a_{1q} \\ ... & ... & ... \\ a_{q1} & ... & a_{qq} \end{vmatrix} c_i = 0 , \qquad i = q+1, ..., m .$$

That is,

$$a_{i1}x_1 + ... + a_{in}x_n = b_i , \qquad i = q+1, ..., m .$$

Therefore the values of $x_1, x_2, ..., x_n$ satisfy the first q equations of system (12.2) as well as all the rest of the equations. This last comment proves the theorem.

The given theorem also suggests the way to solve system (12.2). We should take those q equations whose coefficients form the main determinant and then find these unknowns whose coefficients form this main determinant. The rest of the unknowns can be chosen arbitrarily.

Example 15. We want to find the solution of the system

$$x_1 - 3x_2 + 2x_3 = -5,$$
$$2x_1 + x_2 + 4x_3 = 4,$$
$$x_1 + 4x_2 + 2x_3 = 9,$$
$$5x_1 + 6x_2 + 10x_3 = 17.$$

By the direct calculation we can make sure that the ranks of the coefficient matrix A and the expanded matrix $\bar{A}$ are equal to two, where

$$A = \begin{pmatrix} 1 & -3 & 2 \\ 2 & 1 & 4 \\ 1 & 4 & 2 \\ 5 & 6 & 10 \end{pmatrix},$$

$$\overline{A} = \begin{pmatrix} 1 & -3 & 2 & -5 \\ 2 & 1 & 4 & 4 \\ 1 & 4 & 2 & 9 \\ 5 & 6 & 10 & 17 \end{pmatrix}.$$

So the system has solutions. Solving the second and third equations we find

$$x_1 = 1 - 2x_3, \qquad\qquad x_2 = 2,$$

which is valid for arbitrary x_3.

§13. Solving Homogeneous Linear Systems of Equations

A system of equations with the right hand side equal to zero is called *homogeneous*. The general system of homogenous equations is

$$a_{11}x_1 + \ldots + a_{1n}x_n = 0,$$
$$a_{21}x_1 + \ldots + a_{2n}x_n = 0,$$
$$\ldots\ldots\ldots\ldots\ldots\ldots\ldots\ldots$$
$$a_{n1}x_1 + \ldots + a_{nn}x_n = 0.$$

$$(13.1)$$

As the rank of the expanded matrix for system (13.1) is always equal to the rank of the coefficient matrix, (13.1) is always soluble, i.e. always has a definite solution. This statement is trivial since (13.1) always has a zero solution, i.e. $x_1 = x_2 = \ldots = x_n = 0$.

The existence of non-zero solutions is proved by the next theorem.

Theorem. A homogeneous system has a non-zero solution if and only if the determinant of the system is equal to zero.

Proof. If (13.1) has a non-zero solution then the determinant of the system is equal to zero. Otherwise the zero solution of system (13.1) is unique.

Conversely, if the determinant of the system is equal to zero, then the rank of the coefficient matrix is less than n. Thus at least one of the unknowns x_s can be taken arbitrarily and, in particular, can be chosen to be non-zero. This argument proves the theorem.

We now consider the particular case when the rank of the coefficient matrix of (13.1) is equal to n-1. In this case there is a non-zero determinant of order n-1 which is part of the matrix of coefficients. Suppose that this determinant is formed using the coefficients from the first $(n$-$1)$ rows. Then (13.1) has the solution

$$x_1 = A_{n1}t, \ x_2 = A_{n2}t, \ \ldots, \ x_n = A_{nn}t,$$

$$(13.2)$$

where A_{n1} is the algebraic addition to the elements of the last row of the determinant of (13.1). By virtue of the supposition, there is at least one non-zero coefficient, A_{ns} $(s=1,2,\ldots,n)$. Putting expressions (13.2) into the first $(n$-$1)$ equations of (13.1) we get the identities

$$a_{i1}A_{i1}t + \ldots + a_{in}A_{in}t \equiv 0, \ \ i = 1,\ldots,n-1,$$

which are corollaries of property 12.

Substituting (13.2) into the last equation of (13.1) we get the identity

$$a_{n1}A_{n1}t + \ldots + a_{nn}A_{nn}t \equiv Dt \equiv 0,$$

which holds by virtue of the fact that the determinant of system D is equal to zero. The parameter t in solution (13.2) takes arbitrary values.

Example 16. We find the solution of the homogeneous system

$$2x_1 - x_2 + 3x_3 = 0,$$
$$-x_1 + 2x_2 - x_3 = 0,$$
$$x_1 + x_2 + 2x_3 = 0.$$

The determinant of the system is equal to zero as shown below

$$D = \begin{vmatrix} 2 & -1 & 3 \\ -1 & 2 & -1 \\ 1 & 1 & 2 \end{vmatrix} = 2 \cdot \begin{vmatrix} 2 & -1 \\ 1 & 2 \end{vmatrix} + \begin{vmatrix} -1 & -1 \\ 1 & 2 \end{vmatrix} + 3 \cdot \begin{vmatrix} -1 & 2 \\ 1 & 1 \end{vmatrix} = 0.$$

We then find that the algebraic additions of the elements of the last row are given by

$$A_{31} = \begin{vmatrix} -1 & 3 \\ 2 & -1 \end{vmatrix} = -5, \quad A_{32} = -\begin{vmatrix} 2 & 3 \\ -1 & -1 \end{vmatrix} = -1,$$

$$A_{33} = \begin{vmatrix} 2 & -1 \\ -1 & 2 \end{vmatrix} = 3.$$

Therefore, the system has the solution

$$x_1 = -5t, \quad x_2 = -t, \quad x_3 = 3t,$$

which depends on one parameter.

§14. Other Methods for Solving Linear Systems

To conclude we shall look at some other methods for solving linear equations.

1. **Gauss's Method**

Gauss's method is one of the most convenient methods to solve linear equations. The essence of the method is that by adding linear combinations of the equations to each other, we may bring the system an equivalent system with a special format. Namely that all elements of the associated matrix who lie below the diagonal all equal zero.

Example 17. We solve the following system of equations:

$$x_1 + 2x_2 + 3x_3 = 5,$$
$$2x_1 + 6x_2 + 7x_3 = 10,$$
$$x_1 + 4x_2 + 7x_3 = 11.$$

Subtracting twice the first equation from the second, subtracting the first equation third, and finally subtracting the new second equation from the new third one leads to an equivalent system of equations of the required form

$$x_1 + 2x_2 + 3x_3 = 5,$$

$$2x_2 + x_3 = 0,$$

$$3x_3 = 6.$$

Solving these equations simultaneously we get

$$x_3 = 2, \qquad x_2 = -1, \qquad x_1 = 1.$$

2.   The Sequential Approximation Method

We can write a system of equations in the form

$$x_1 + a_{11}x_1 + \ldots + a_{1n}x_n = b_1,$$

$$x_2 + a_{21}x_1 + \ldots + a_{2n}x_n = b_2,$$

$$\ldots\ldots\ldots\ldots\ldots\ldots\ldots\ldots\ldots\ldots\ldots\ldots$$

$$x_n + a_{n1}x_1 + \ldots + a_{nn}x_n = b_n,$$

$$(14.1)$$

where the norms of elements a_{is} are small enough. We shall introduce the following definitions

$$X = \begin{pmatrix} x_1 \\ x_2 \\ \ldots \\ x_n \end{pmatrix}, \qquad B = \begin{pmatrix} b_1 \\ b_2 \\ \ldots \\ b_n \end{pmatrix},$$

$$A = -\begin{pmatrix} a_{11} & a_{12} & \ldots & a_{1n} \\ a_{21} & a_{22} & \ldots & a_{2n} \\ \ldots & \ldots & \ldots & \ldots \\ a_{n1} & a_{n2} & \ldots & a_{nn} \end{pmatrix}.$$

So we can write (14.1) in a vector form

$$X - AX = B.$$

(14.2)

We now create the sequence of vectors X_s $(s=0,1,2,...)$; $X_0=0$, $X_{s+1}=B+AX_s$ $(s=0,1,2,...)$. Iterating this for all s we have

$$X_1 = B, \quad X_2 = B + AB, \quad X_3 = B + AB + A^2B, \dots .$$

For the sequence X_s to converge to a solution X it is sufficient that the norm of matrix A is smaller than one. The norm of matrix A is a non-negative real number $|A|$ and one of the possible definitions is expressed by the formula

$$|A| = \max_i \sum_{s=1}^{n} |a_{is}| .$$

(14.3)

Example 18. We solve the following system:

$$10x - y = 1,$$
$$x + 10y = 2.$$

By dividing the equations by 10 and representing them in a more convenient fashion for the application of the method of sequential approximations we get

$$x = 0.1 + 0.1y,$$
$$y = 0.2 - 0.1x.$$

Suppose $x_0=y_0=0$. We shall define sequences x_s, y_s as

$$x_{s+1} = 0.1 + 0.1y_s,$$
$$y_{s+1} = 0.2 - 0.1x_s.$$

We have the sequence

$$x_1 = 0.1, \qquad x_2 = 0.12, \qquad x_3 = 0.119,$$
$$y_1 = 0.2, \qquad y_2 = 0.19, \qquad y_3 = 0.188,$$
$$x_4 = 0.1188,$$
$$y_4 = 0.1881,$$

which converges to the following system:

$$x = 0.118811881188,$$
$$y = 0.188118811881.$$

Note 6. The method of sequential approximations is *not* universal; nevertheless it is often used for solving infinite linear systems of equations of the type

$$x_i + \sum_{s=1}^{\infty} a_{is} x_s = b_i, \quad i = 1,...,$$

$$(14.4)$$

which satisfy the following condition

$$\sum_{s=1}^{\infty} |a_{is}| \le \theta < 1,$$

$$(14.5)$$

which is equivalent to the condition

$$|A| \le \theta < 1.$$

In the general case for the application of this method for solving system (14.2) it is enough that all eigenvalues of the matrix A are a modulo smaller than one. We remind the reader that the eigenvalues of matrix A are the roots of the equation

$$\det(A - zI) = 0.$$

3. Reduction

While solving a system of equations it is sometimes possible to lower the order of the system to be solved. Consider the system of equations

$$AX + BY = G,$$
$$CX + DY = H,$$

$$(14.6)$$

where X, Y are the unknown vectors of order n and m respectively. A, B, C, D are known matrices. G, H are known vectors. Let D be invertible, then the second system can be solved for Y such as

$$Y = D^{-1}H - D^{-1}CX,$$

where D^{-1} is the inverse of D. Substituting this solution into the first system we find a system of order n.

$$\left(A - BD^{-1}C\right)X = G - BD^{-1}H.$$

$$(14.7)$$

The method of reduction is applied to infinite linear systems of equations. For example, if condition (14.5) is satisfied only for $i>N$, then the initial system (14.4) is solved for $x_{i+1}, x_{i+2}, \ldots$ given $x_1, \ldots x_i$. The system can be solved by virtue of condition (14.5). Excluding the unknowns $x_{i+1}, x_{i+2}, \ldots$ from the first i equations we find a system of N equations with N unknowns $x_1, \ldots, x_N$. Let there be a non-zero solution to the infinite linear system of equations

$$\begin{matrix} AX + BY = 0 \\ CX + DY = 0 \end{matrix}, \qquad R = \begin{pmatrix} A & B \\ C & D \end{pmatrix},$$

$$(14.8)$$

where A is an $n \times n$ matrix and D is an infinite matrix which satisfies the condition

$$|D - I| < 1,$$

where I is the infinite identity matrix. In this case there is an inverse matrix D^{-1} such as

$$D^{-1} = \left(I + (D - I)\right)^{-1} = I + (I - D) + (I - D)^2 + (I - D)^3 + \ldots$$

.

Reducing system (14.8) by formula (14.7) we find an existence condition for a non-zero solution, namely

$$\det\left(A - BD^{-1}C\right) = 0.$$

(14.9)

We can arrive at equation (14.9) using the following method. Take the infinite auxiliary matrix

$$Q = \begin{pmatrix} I_n & 0 \\ C & D \end{pmatrix},$$

$$Q^{-1} = \begin{pmatrix} I_n & 0 \\ -D^{-1}C & D^{-1} \end{pmatrix}.$$

We transform the system as follows:

$$\det R = \det\left(RQ^{-1}\right)\det Q = \det\begin{pmatrix} A - BD^{-1}C & BD^{-1} \\ 0 & E \end{pmatrix},$$

$$\det\begin{pmatrix} I_n & 0 \\ C & D \end{pmatrix} = \det\left(A - BD^{-1}C\right)\det D.$$

As $\det D \neq 0$, $\det R = 0$ takes the form of equation (14.9). Applying the method for constructing matrix Q we can use the known expansion

$$Q^{-1} = \left(I + (Q - I)\right)^{-1} = I + (I - Q) +$$
$$+ (I - Q)^2 + (I - Q)^3 + \dots.$$

This method proved efficient in reducing Hill's infinite determinant to the finite determinant.

4.   The Combined Method

There is a system solution of the form (14.6) where $\det A \neq 0$, $\det D \neq 0$ and the norms of matrices B and C are small enough. We shall estimate matrices A^{-1} and D^{-1} and instead of system (14.6), we shall solve the modified system

$$\begin{pmatrix} A^{-1} & 0 \\ 0 & D^{-1} \end{pmatrix}\begin{pmatrix} A & B \\ C & D \end{pmatrix}\begin{pmatrix} X \\ Y \end{pmatrix} = \begin{pmatrix} A^{-1} & 0 \\ 0 & D^{-1} \end{pmatrix}\begin{pmatrix} G \\ H \end{pmatrix},$$

which can be written as

$$X = A^{-1}G - A^{-1}BY,$$
$$Y = D^{-1}H - D^{-1}CX,$$

which is convenient for applying the method of sequential approximations. This method can be applied for finding a inverse matrix. One can save time expended in calculation by dealing with matrices of an order lower than the system.

Bibliography:

1. Grassmann H. Die Ausdehnungslehre, 1844, 2 publ. 1861
2. Koch H. Ueber das Nichtverschwinden einer Determinante nebst Bemerkungen über Systeme unendlich vieler linearer Gleichungen, Deutsche Math. Verl. 22.285-291, 1913
3. Klein F. Vorlesungen über die Entwicklung der Mathematik im 19. Jahrhundert. Teil 1. Berlin, Verlag von Julius Springer 1926
4. Автоматизация проектирования устройств измерительной техники / Ю. М. Туз, А. И. Забарный, Б. Н. Белоусов и др. – К.: Выща шк. Головное изд-во 1988. – 288с., 142 ил. – Библиогр.: 25 назв.
5. Математика. Числовые системы, многочлены, определители / Т. Г. Стрижак, Ю. М. Туз - К.: УМК ВО 1991. – 436 с.

Prepared by
Benedikt Rudolph,
a trainee student IASTE
from TU Munich, Germany

Instead of the afterword

Why has Grassmann Algebra (Hermann Grassmann was born in 1809) been included in the project "Modern Mathematics for Engineers"? (The project was launched in 2008).

The answer was given by French mathematician Jean Alexandre Eugène Dieudonné (born in 1906): "You should never say that some part of mathematics died, because it can be said today, but tomorrow someone will deal with this theory and bring something new into it and it will be brought back to life".

As a matter of fact Professor NTU KPI, State Award Laureate in Science and Technique Tuz Yu. used Grassmann Algebra while creating Volt Standard for analysing measurement inaccuracy in 2008 and on its basis created the "Trajectory Method", an analogue of Taylor series.

Since Grassmann is not very well-known for mathematicians, he shall be introduced in the following passage.

Hermann Guenther Grassmann was born in Stettin (today Szczecin, Poland) on April 15, 1809. At that time Stettin was occupied by the French troups under Napoleon I. Hermann's father Justus Grassmann joined the Prussian army in 1813 while Hermann's mother Johanna Grassmann fled with the four children. Stettin was besieged and occupied by the Prussians. Thus in 1814 Justus Grassmann could pick up his former work as a teacher at the local Gymnasium and his family returned to Stettin.

In contrary to other great scientists Hermann Grassmann did not appear to be a genius in his youth. However he worked hard to pass his final school exams with best grades in 1827. Together with his older brother Gustav he moved to Berlin afterwards to study theology. Besides he attended lectures in psychology whereas during his studies he never joined any mathematical classes. He returned to Stettin in 1830 and as an autodidact he taught himself philology, mathematics and physics. From 1831 on he worked as a teacher. In the following years he passed exams in theology, mathematics and physics. Furthermore he published two mathematical works.

His main work, the *Ausdehnungslehre* (theory of extension), which contained his new vector theory, was published in 1844. However, due to the way of presentation, the work was not accepted by experts of that time. The philosophical foundation, the self-created terminology and the lack of formulas did not match their expecation of mathematical works. This led

Grassmann into a struggle for acceptance. In the following years he published works that were based on the *Ausdehnungslehre* and joined competitions to promote his work.

In 1849 Grassmann married Marie Therese Knappe. They had eleven children.

In 1862 Grassmann published a new edition of the *Ausdehnungslehre* in which he completely changed the way of presentation. However, he was forced to wait until the late 1860s to see his mathematical achievements finally accepted among experts. This can be considered the reason why in these years he did not publish any more mathematical works but concentrated on philology.

Hermann Guenther Grassman died in September 26, 1877.

Source: Petsche, Hans Joachim *Graßmann*, Birkhäuser Verlag Basel, 2006

Summary of results from the first volume

Minimax criteria
of stability

$$L = T - f$$

$$\max_{q} \left\langle \min_{\dot{q}} L\left(t, \dot{q}_j, q_j\right) \right\rangle$$

"Mathematical Truth stays put in centuries,
Metaphysical Ghosts disappear
like delirium of the sick."
Voltaire

Minimax Criterion of Stability is the first volume within the project *Modern Mathematics for Engineers*.
The project *Modern Mathematics for Engineers* was launched by   AUS–DAAD and the National

Committee IAESTE Ukraine. The purpose of the project is to publish the most essential mathematical results, to present them with enough clarity for engineers to apply and, ante omnia, to realize the scientific exchange in the sphere of applied mathematics, the oscillation theory, theoretical mechanics, etc.

We expect to realize this purpose first of all with the help of IAESTE trainee students.

We invite for cooperation all mathematicians and engineers who are interested in having their scientific works published in the frame of the project.

We were inspired to launch the project by the positive experience of the researchers of the Californian University, who published the monograph *Modern Mathematics for Engineers*[1] more than half a century ago, in 1956. This work was very successful. As a matter of fact this monograph laid the solid foundation for the applied mathematics to develop successfully and steadily. This monograph is an excellent example to follow. Thus, like 50 years ago, we are launching the project with researches dedicated to the pendulum. We can hardly remember any other

---

[1]   Edited by Edwin F. Beckenbach (Professor of Mathematics, University of California, Los Angeles). New York, Toronto, London: McGraw-Hill 1956.

mechanism which is simpler than the pendulum, the mechanism whose scientific life has been so rich in application in different spheres of organic and inorganic nature, as the pendulum has lived a long and rich in discoveries life. I think it is the right time to build a monument to a simple and meaningful device.

Our contribution into realization of this project included taking the following steps:

1) publishing the monograph *Research Methods of the Pendulum Dynamic Systems* [4];

2) receiving "Minimax Criterion of Stability" and applying it in order to research a number of mathematical models and proved, in particular, that any position of the pendulum, even a horizontal one in the vertical plane, can be made stable with the help of the suspension point oscillations;

3) building an installation to demonstrate our theoretical results.

All in all we tried our best to state the results avoiding proving theorems as their translation from Russian into English might have some inadequacy as grammars of these languages have a certain level of ambiguity. Instead of proving the theorems we shall follow Lopital's words, the author of the first textbook on Mathematical Analysis, and say, "We pass the word of honor that the theorem is true".

b-site: http://www.iaeste.org.ua

Oscillations – what are they?

"Nature prefers oscillating motions
in all demonstrations of life.
Not without reason we can assume that there is some
optimality property at the back
of this phenomenon."[8]

It would be rather difficult to mention all publications about the Pendulum. As a matter of fact, oscillations of pendulum systems were researched by Galileo, Newton, Euler, Huygens and many other scientists who made a great contribution into studying the mechanism of the pendulum motion, which was of great importance for the history of mankind's discoveries and technical progress. The pendulum has been used in clocks to define time, in special devices to measure the terrestrial gravitation, as a plumb line in building to define the vertical, etc. The Earth's rotation was proved with the help of the pendulum oscillations as well.

Pendulum systems are, as a rule, non-linear and require specific methods of research. For instance, creation of the elliptical function theory by Abel, Jacoby, and Weierstrass is also connected with the research of the mathematical pendulum oscillations.

Nowadays due to the mathematicization of research in different sciences there has been an increasing interest in studying motions of pendulum systems. The following special terms have appeared: the pendulum law of the population migration, the pendulum law of the rhythm regulator action, the pendulum of emotions, etc.

It is a well-known fact that all living organisms have so called biological clocks, at the basis of which there is an oscillating system – a non-linear oscillator. It is well known that the vestibular apparatus of animals and humans contains three non-linear pendulum systems, which are located in three mutually perpendicular planes. Oscillations are present everywhere: in the opening of a flower at the sunrise, in the growth of an embryo, in germinating of a grain, in the heart beating, in the work of a dental drill and a jackhammer, in the rise and fall of the tide. Besides, wherever there is life, there are oscillations at the cellular level.

The nature of forces which cause oscillations is variable, but the result of these forces is the same, namely - oscillations.

This information bears a descriptive character as we would like to attract the readers' attention to the oscillation theory to convince them of the fact that oscillation processes imply something much deeper than what we actually know: that the source of the oscillating processes is a hidden potential force and kinetic energy.

Oscillations of a non-linear oscillator can be forced or can have a free-running character (such as the heart or aorta beating, biorhythms, etc.). In fact, the pendulum motion laws, periodic or almost-periodic oscillations, are immanent in the whole physical world.

A human being receives the main information about the outer world via sound and light oscillations, analyzing which scientists use pendulum systems. The latter are found in different engineering tasks. For example, oscillations in electrical and mechanical systems, rocking of ships on water, oscillations of satellites, vibration of the hull and the wings of planes, movements of cables, chains, travelling or gantry cranes, etc. All these phenomena are explained with analogous differential equations which describe motions.

At this point our introduction is completed and we turn to the beautiful algorithmic mathematical language: *"A" is given, "B" is to be proved.*

Minimax criteria of stability

The stability of oscillations of a mechanical or electrical system, which is affected by low amplitude high-frequency disturbances, is researched. It is shown that the affect of vibration can be replaced by the action of some potential force. Under the action of some disturbance the stable equilibrium position can become unstable, and the unstable equilibrium position can become stable. New positions of stable dynamic equilibrium can also appear. Conclusions about the stability can be made with the help of the minimax criteria for stability described in this work. The criteria for stability rests on the time averaging operation of the canonical system of differential equations. The minimax criteria for stability is applied when the frequency of external disturbance is substantially higher than the natural vibration of the quiescent system.

In this work the minimax criteria for stability is applied in the research of the stability of the pendulum systems equilibrium position. In particular, it is found that any position of the pendulum, even a horizontal one, can be made stable via vibrations of the pendulum's suspension point.

An experimental installation to check the received theoretical conclusions was created. All the conclusions were proved correct.

This work illustrates simple examples of the minimax criteria for stability application. It also describes the experiment. At the end of this work the mathematical survey of the minimax criteria for stability is presented.

§1   Minimax stability criteria definition

Here the movement of a mechanical or electrical system is studied. We shall define the generalized coordinates by q_j $(j = 1,...,n)$, the generalized velocities by $\dot{q}_j$ $(j = 1,...,n)$ and the time by t.

Let $L = L(t,\dot{q}_j,q_j)$ be the kinetic potential, where $L = T - \Pi$, and T is the kinetic energy of the system and Π is the potential energy. We shall find the minimum of the kinetic potential L as a function of the generalized velocities $\dot{q}_j$,

$$\Pi_0(t,q_j) = \min_{\dot{q}_s} L(t,\dot{q}_j,q_j), \quad (j,s = 1,...,n).$$

(1)

We shall exclude from $L(t,\dot{q}_j,q_j)$ the derivative $\dot{q}_j$ by the necessary conditions of extremum

$$\frac{\partial L(t,\dot{q}_j,q_j)}{\partial \dot{q}_s} = 0, \quad (j,s = 1,...,n).$$

(2)

Using the operation of time averaging t on the function $\Pi_0(t, q_j)$ we are able to eliminate t

$$\Pi_0(q_j) \equiv \langle \Pi_0(t, q_j) \rangle \equiv \lim_{T \to +\infty} \frac{1}{T} \int_0^T \Pi_0(t, q_j) dt .$$

(3)

If the function $\Pi_0(q_j)$ has a maximum at the point $q_j = q_{j0}$ $(j = 1, ..., n)$, then the maximum corresponds to the stable position of the dynamic equilibrium of the initial oscillating system. Meanwhile the system is steadily oscillating at the position $q_j = q_{j0}$ $(j = 1, ..., n)$.

To find the position of the stable state it is first necessary to find the minimum of the function $L(t, \dot{q}_j, q_j)$ via variable $\dot{q}_j$, and then find the maximum of the function using,

$$\max_{q_j} \left\langle \min_{\dot{q}_j} L(t, \dot{q}_j, q_j) \right\rangle ,$$

by which the name *'minimax criteria of stability'* arises.

It is important to mention that the conditions of stability are found without using Lagrange's differential equations [1]

$$\frac{d}{dt}\frac{\partial L}{\partial \dot{q}_s} - \frac{\partial L}{\partial q_s} = 0, \quad (s = 1,...,n)$$

and only the knowledge of Lagrange's function is used $L = L\left(t, \dot{q}_j, q_j\right)$.

§2 Stability of the mathematical pendulum oscillations at the upper equilibrium position

We shall consider oscillations at the upper equilibrium position of the mathematical pendulum with length l and mass m, with the vibrating vertical point of support A. We shall define the coordinates of the centre of gravity of the pendulum by x, y, the pendulum angle from the vertical by φ and the deflection of the point A of the pendulum support in a vertical direction (figure 1) by $r(t) = a \sin \omega t$, where a is the amplitude and ω is the frequency of the vibrations.

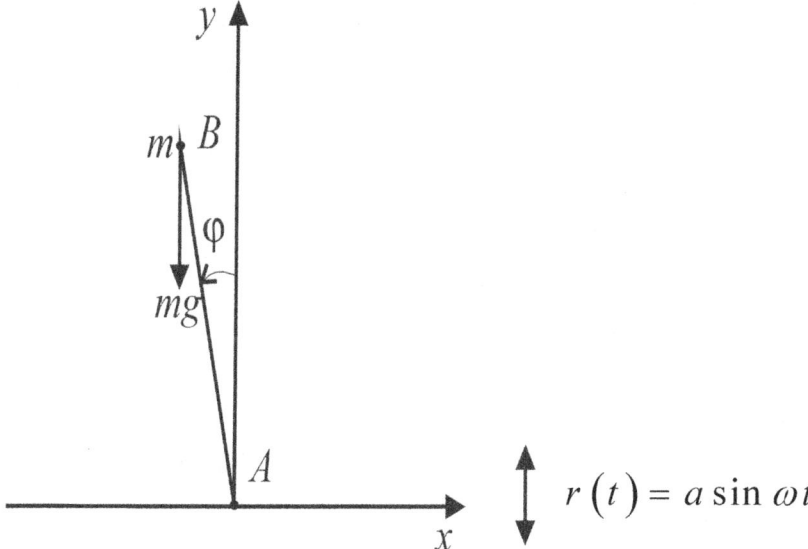

Figure 1

For the coordinates of the centre of gravity we find the following,

$$x = -l\sin\varphi, \quad y = l\cos\varphi + r,$$

and for the velocities

$$\dot{x} = -l\cos\varphi \cdot \dot{\varphi}, \quad \dot{y} = -l\sin\varphi \cdot \dot{\varphi} + \dot{r}.$$

The kinetic and potential energies are found to have the following forms

$$T = \frac{m}{2}\left(\dot{x}^2 + \dot{y}^2\right) = \frac{m}{2}\left(l^2\dot{\varphi}^2 - 2l\dot{\varphi}\sin\varphi\cdot\dot{r} + \dot{r}^2\right),$$

$$\Pi = mg\left(l\cos\varphi + r\right).$$

Then the Lagrange function is as follows,

$$L = T - \Pi = \frac{m}{2}\left(\left(l\dot{\varphi} - \dot{r}\sin\varphi\right)^2 + \dot{r}^2\cos^2\varphi\right) - mg\left(l\cos\varphi + r\right)$$

(4)

and $\min\limits_{\dot{\varphi}} L$ is found with the help of the equation

$$\frac{\partial L}{\partial \dot{\varphi}} = m\left(l\dot{\varphi} - \dot{r}\sin\varphi\right)\cdot l = 0$$

to give the following expression

$$\min\limits_{\dot{\varphi}} L = \frac{m}{2}\dot{r}^2\cos^2\varphi - mgl\cos\varphi - mgr$$

(5)

As $r(t) = a\sin\omega t$ we get

$$\langle r \rangle = \langle a \sin \omega t \rangle = 0, \qquad \langle \dot{r}^2 \rangle = \langle a^2 \omega^2 \cos^2 \omega t \rangle = \frac{a^2 \omega^2}{2}$$

where <X> is the time average of the function X.

Therefore the following is found,

$$\Pi_0(\varphi) = \langle \min_{\dot{\varphi}} L \rangle = mgl \cos\varphi - \frac{m}{2} \frac{a^2 \omega^2}{2} \cos^2 \varphi,$$

where $-\Pi_0(\varphi)$ is a dynamic analogue of potential energy.

The function $-\Pi_0(\varphi)$ has its minimum at the point $\varphi = 0$, if the following condition is satisfied

$$\frac{\partial^2 \Pi_0(\varphi)}{\partial \varphi^2} < 0 \quad \text{or} \quad mgl - \frac{m}{2} a^2 \omega^2 < 0.$$

The stability criteria is reduced to the following inequality,

$$a^2 \omega^2 > 2gl.$$

(6)

Relation (6) was discovered earlier in relevant scientific work. It is obvious that it was first discovered in some works [2, 3], and later repeated in the works of N. Bogolyubov, P. Kapitsa, G. Stoker, K.Valeev, T. Stryzhak.

Summary of results from the second volume

Tamara G. Stryzhak

Frequency Criterion of Stability

The Perseval equality:

$$\int_0^\infty y^2(t)\,dt = \frac{1}{2\pi} \int_{-\infty}^\infty \left| f(i\omega) \right|^2 d\omega$$

There are different criteria for the stability of the solution of linear differential equations, which are based on the characteristic equation analysis. In this work the reader will find principally new ways of frequency criterion for stability derivation, which are based on Perceval formula and resolution of operator equations in Banach space. At first we consider simple examples and then a more general theory will be stated. Any remarks and recommendations will be appreciated and taken into considerations in the following issues.

Abstract from the next volume

The Principle of Reduction

The principle of reduction is a common idea which states that while researching the stability of solutions of the dynamic system (the system of differential, difference, differential-difference equations) the order of the researched system can be decreased. As the main obstacle in researching a dynamic system is a large dimension, then decreasing the order essentially simplifies the stability researching process.

This work contains some ways of decreasing the order of the system. We do not give a detailed study in order not to hamper the understanding of the main ideas, and the rigour of explanations is replaced by examples and references to the original works.

This work will be published in two languages: English and Russian.

ibidem-Verlag

Melchiorstr. 15

D-70439 Stuttgart

info@ibidem-verlag.de

www.ibidem-verlag.de
www.ibidem.eu
www.edition-noema.de
www.autorenbetreuung.de

Made in the USA
Monee, IL
11 September 2025

25505497R00069